Edited by
Hans-Jörg Bart
and Stephan Pilz

**Industrial Scale Natural
Products Extraction**

Related Titles

Soetaert, W., Vandamme, E. J. (eds.)

Industrial Biotechnology

Sustainable Growth and
Economic Success

2010
ISBN: 978-3-527-31442-3

Buchholz, K., Collins, J.

Concepts in Biotechnology

History, Science and Business

2010
ISBN: 978-3-527-31766-0

Katoh, S., Yoshida, F.

Biochemical Engineering

A Textbook for Engineers,
Chemists and Biologists

2009
ISBN: 978-3-527-32536-8

Dunn, P., Wells, A., Williams, M. T. (eds.)

Green Chemistry in the Pharmaceutical Industry

2009
ISBN: 978-3-527-32418-7

Edited by Hans-Jörg Bart and Stephan Pilz

Industrial Scale Natural Products Extraction

WILEY-VCH Verlag GmbH & Co. KGaA

The Editors

Prof. Dipl.-Ing. Dr. Hans-Jörg Bart
Technical University of Kaiserslautern
Lehrstuhl für Thermische
Verfahrenstechnik
Gottlieb-Daimler-Straße 44
67663 Kaiserslautern
Germany

Dr. Stephan Pilz
Evonik Degussa GmbH
Rodenbacher Chaussee 4
63457 Hanau-Wolfgang
Germany

■ All books published by **Wiley-VCH** are carefully produced. Nevertheless, authors, editors, and publisher do not warrant the information contained in these books, including this book, to be free of errors. Readers are advised to keep in mind that statements, data, illustrations, procedural details or other items may inadvertently be inaccurate.

Library of Congress Card No.: applied for

British Library Cataloguing-in-Publication Data
A catalogue record for this book is available from the British Library.

Bibliographic information published by the Deutsche Nationalbibliothek
The Deutsche Nationalbibliothek lists this publication in the Deutsche Nationalbibliografie; detailed bibliographic data are available on the Internet at <http://dnb.d-nb.de>.

© 2011 Wiley-VCH Verlag & Co. KGaA, Boschstr. 12, 69469 Weinheim, Germany

All rights reserved (including those of translation into other languages). No part of this book may be reproduced in any form – by photoprinting, microfilm, or any other means – nor transmitted or translated into a machine language without written permission from the publishers. Registered names, trademarks, etc. used in this book, even when not specifically marked as such, are not to be considered unprotected by law.

Composition Toppan Best-set Premedia Ltd., Hong Kong
Printing and Binding Strauss GmbH, Mörlenbach
Cover Design Formgeber, Eppelheim

Printed in the Federal Republic of Germany
Printed on acid-free paper

ISBN: 978-3-527-32504-7

Contents

Preface *XIII*
List of Contributors *XV*

1 **Extraction of Natural Products from Plants – An Introduction** *1*
Hans-Jörg Bart
1.1 Introduction *1*
1.2 Cultivation *6*
1.3 Extraction *8*
1.3.1 Solvents *8*
1.4 Extraction Techniques *15*
1.5 Purification *17*
1.5.1 Chromatography *17*
1.5.1.1 Adsorption Chromatography *17*
1.5.1.2 Partition Chromatography *18*
1.5.1.3 Ion Exchange Chromatography *18*
1.5.1.4 Gel Chromatography *19*
1.5.1.5 (Bio-) Affinity Chromatography *19*
1.5.2 Continuous Techniques *19*
1.5.2.1 True Moving Bed (TMB) Chromatography *20*
1.5.2.2 Simulated Moving Bed (SMB) Chromatography *20*
1.5.2.3 Annular Chromatography *22*
1.5.2.4 Carrousel Adsorbers *22*
References *24*

2 **Solubility of Complex Natural and Pharmaceutical Substances** *27*
Feelly Ruether and Gabriele Sadowski
2.1 Introduction *27*
2.2 Solubility Calculations *27*
2.2.1 Solubility of a Pure Solute in Solvents and Solvent Mixtures *27*
2.2.2 pH-Dependence of Solubility *29*
2.2.3 Solubility of Racemic Compounds *29*
2.3 Thermodynamic Modeling *30*

2.3.1	PC-SAFT Equation of State 30
2.3.1.1	Hard-Chain Contribution A^{hc} 30
2.3.1.2	Dispersion Contribution A^{disp} 31
2.3.1.3	Association Contribution A^{assoc} 31
2.3.2	Estimation of PC-SAFT Parameters 33
2.4	Examples 35
2.4.1	Solubility of Estriol, Estrone, and Sitosterol in Different Solvents 35
2.4.2	Solubility of Beta-carotene in Supercritical Carbon Dioxide 37
2.4.3	Solubility of Paracetamol in Pure Solvents and Solvent Mixtures 40
2.4.4	Solubility of DL-Methionine as Function of pH 44
2.4.5	Solubility of Mandelic Acid Enantiomers and Racemic Mandelic Acid in Water 45
2.5	Summary 49
	Symbols 49
	Latin Symbols 49
	Greek Symbols and Special Characters 50
	Superscripts 50
	Subscripts 50
	Appendix 51
	Hard-chain Reference Contribution 51
	Dispersion Contribution 52
	Association Contribution 53
	References 53

3	**Alternative Solvents in Plant Extraction** 55
	Volkmar Jordan and Ulrich Müller
3.1	Introduction 55
3.2	Ionic Liquids in the Extraction of Natural Compounds from Plant and Fungi 56
3.2.1	Characteristics of Ionic Liquids 57
3.2.1.1	Physicochemical Properties 57
3.2.1.2	Environmental and Safety Aspects 60
3.2.2	Application of Ionic Liquids in Plant Extraction 61
3.2.2.1	Application of ILs in Extraction 61
3.2.2.2	Removal of Target Substance from Extract and Separation of Solvent from Spent Biomass 61
3.2.2.3	Example 1: Extraction of Artemisinin 62
3.2.2.4	Example 2: Extraction of Lignin 64
3.3	Surfactants and Aqueous Two-Phase Systems in Plant Extraction 65
3.3.1	Characteristics of Surfactant–Water Mixtures 67
3.3.2	Behavior of Nonionic Surfactants in Aqueous Solution 71
3.3.3	Micellar Extraction and Cloud Point Extraction 73
3.3.4	Reversed Micellar Extraction 74
3.3.5	Equilibrium Partition of Target Substances in Aqueous Surfactant Solutions 75

3.3.6	Examples for the Use of Surfactants in Plant Extraction	77
3.3.6.1	Plant Extraction Using Micellar and Cloud Point Extraction	77
3.3.6.2	Plant Extraction Using Reverse Micelles	80
3.4	Summary	83
	References	84

4   High Pressure Processing   87
Rudolf Eggers and Stephan Pilz
4.1   Introduction   87
4.2   Supercritical Fluids   90
4.2.1   General   91
4.2.2   Physical Properties   93
4.2.3   Solvent Power and Solubility   94
4.3   Physical Properties – Mass Transfer Characteristics   95
4.4   Process Units   99
4.4.1   Pre-Treatment   99
4.4.1.1   Mechanical Pre-treatment   100
4.4.1.2   Thermal and Hydrothermal Treatment   103
4.4.1.3   Innovative Methods   104
4.4.2   Extraction   104
4.4.3   Separation   106
4.4.4   Post-Treatment   107
4.4.4.1   Degassing   107
4.4.4.2   Further Separation   107
4.4.4.3   Stabilization   107
4.4.4.4   Formulation   108
4.5   Process Design and Operation   108
4.5.1   Process Concept   108
4.5.2   Pressurization and Depressurization   109
4.5.3 CO_2 Recycle Loop   110
4.5.4   Scale-Up   111
4.5.5   Costs   113
4.6   Applications   113
          Acknowledgments   120
          References   120

5   Process Engineering and Mini-Plant Technology   123
Jochen Strube, Werner Bäcker and Michael Schulte
5.1   Introduction   123
5.2   Chromatographic Screening   132
5.3   Preparative Task   134
5.3.1   Crude Mixture Pre-Treatment   137
5.3.2   Final Purification of the End Product   139
5.3.2.1   Batch Chromatography   140
5.3.2.2   Annular Chromatography   140

5.3.2.3 Steady-State Recycling Chromatography *141*
5.3.2.4 Simulated Moving Bed (SMB) Chromatography *142*
5.4 Liquid–Liquid Extraction *143*
5.4.1 Introduction *143*
5.4.2 Choice of Extracting Agent *144*
5.4.2.1 Selection Criteria *145*
5.4.3 Data of Chemical and Physical Properties *146*
5.4.4 Specification of Separation Sequence *147*
5.4.5 Operation Concepts in Extraction *148*
5.4.6 Determination of Equipment Efforts *151*
5.5 Mini-Plant Technology for Extraction Process Development *155*
5.6 Cost Estimation *159*
5.7 Total Process Development *160*
5.7.1 Mini-Plant Technology *160*
5.7.2 Examples of Typical Processes *164*
5.7.2.1 Quassia Extracts *164*
5.7.2.2 Spruce Bark Extracts *165*
5.7.2.3 Pepper Extracts *167*
5.7.2.4 Vanilla Extracts *167*
5.7.2.5 Taxol Extracts *168*
5.7.3 Process Modeling and Scale-Up *171*
5.8 Future Developments *175*
Acknowledgments *175*
References *176*

6 Extraction Technology *181*
Andreas Pfennig, Dirk Delinski, Wilhelm Johannisbauer and Horst Josten
6.1 Introduction *181*
6.2 Extraction Process Basics *181*
6.2.1 Introduction *181*
6.2.2 Extraction Techniques *183*
6.2.2.1 Solid–Liquid Extraction *183*
6.2.2.2 Process Alternatives *185*
6.2.2.3 Continuous or Discontinuous Extraction *186*
6.2.3 Further Considerations *186*
6.3 Experimental Procedures *187*
6.3.1 Extraction Curves *187*
6.3.2 Determination of Equilibrium and Kinetics *187*
6.3.3 Standard Laboratory Experiment *190*
6.3.4 Example Extraction of Pepper *191*
6.3.4.1 Calibration with Pure Component *191*
6.3.4.2 Pre-treatment of Solid Raw Material *191*
6.3.4.3 Extraction Kinetics *192*
6.4 Theoretical Modeling and Scale-Up *193*

6.4.1	Macro-Model	194
6.4.1.1	Ideal Stirred Tank	194
6.4.1.2	Ideal Plug-Flow Extractor	195
6.4.1.3	Other Macro-Models	195
6.4.2	Micro-Models	196
6.4.2.1	Other Micro-Models	197
6.4.3	Fitting to Experimental Data and Model-Based Experimental Design	199
6.4.4	Scale-Up Design	199
6.5	Industrial Extraction Equipment	200
6.5.1	Pre-processing of Plant Material	200
6.5.2	Selection of Extraction Solvents	201
6.5.3	Solid–Liquid Extraction Processes	202
6.5.3.1	Batch Processes	202
6.5.3.2	Continuous Processes	204
6.5.4	Extractor Design	205
6.5.4.1	Batch Equipment	206
6.5.4.2	Continuous Extractors	208
6.5.5	Future Developments	212
6.5.6	Extract Treatment	215
6.5.7	Depleted Plant Treatment	216
	Symbols	217
	Latin Symbols	217
	Greek Symbols and Special Characters	218
	Subscripts	218
	References	218

7 Extraction of Lignocellulose and Algae for the Production of Bulk and Fine Chemicals 221
Thomas Hahn, Svenja Kelly, Kai Muffler, Nils Tippkötter and Roland Ulber

7.1	Introduction	221
7.2	Products from Lignocellulose	222
7.2.1	Cellulose Fraction	222
7.2.2	Hemicellulose Fraction	224
7.2.3	Extraction of Fermentation Inhibitors	225
7.2.4	Lignin Fraction	226
7.2.5	Continuous Extraction of Lignocellulose	227
7.3	Polysaccharides and Sporopollenin from Marine Algae	227
7.3.1	Agar	227
7.3.2	Carrageenan	230
7.3.3	Alginic Acid	232
7.3.4	Fucoidan	235
7.3.5	Sporopollenin/Algaenan	238
	References	242

8	**Natural Products – Market Development and Potentials** 247
	Sybille Buchwald-Werner and Friedrich Bischoff
8.1	Introduction 247
8.2	Natural Product Market Overview 247
8.2.1	Market for Extracts 248
8.3	Natural Products in Food and Functional Food 249
8.3.1	Functional Food Concept Based on Natural Products 249
8.3.2	Natural Product Development 250
8.3.3	Food Applications of Natural Products 252
8.3.4	Success Story Plant Sterols for Heart Health 256
8.3.5	Baobab – The New Superfruit 257
8.4	Natural Products for Pharmaceutical Applications 259
8.4.1	Existing Products and Innovation 259
8.4.2	Terpenes: Paclitaxel (Taxol) – The Posterchild 262
8.4.3	Alkaloids: A Group of Potent APIs 263
8.4.3.1	Camptothecin-derived APIs 263
8.4.3.2	Omacetaxine (Homoharringtonine) 264
8.4.3.3	Opiate Alkaloids 264
8.5	Summary 266
	References 267

9	**Regulations and Requirements** 269
	Guido Ritter and Alwine Kraatz
9.1	Introduction 269
9.2	Definition of Plant Extracts as Medicines or Foods – A Borderline Issue 270
9.2.1	Medicines 270
9.2.2	Foods 271
9.3	Application of Plant Extracts in Flavorings, Functional Foods, Novel Foods and Food Supplements 272
9.3.1	Flavorings 272
9.3.2	Functional Foods 273
9.3.3	Regulation on Nutrition and Health Claims Made on Foods 273
9.3.3.1	Nutrition Claim 275
9.3.3.2	Health Claim 275
9.3.3.3	Reduction of Disease Risk Claim 276
9.3.4	Regulation Concerning Novel Foods and Novel Food Ingredients 276
9.3.5	Regulation Concerning Food Supplements 277
9.3.6	The Health Benefits of Food Products Containing Plant Extracts 278
9.3.6.1	Mode of Function Plant Extracts as Antioxidants 278
9.3.7	Examples of Applications of Plant Extracts in Functional Foods and Dietary Supplements 278
9.3.7.1	Black and Green Tea 279
9.3.7.2	Gingko Biloba 280
9.3.7.3	Soy Protein 280

9.4	Requirements Concerning Quality and Safety Issues	*281*
9.4.1	Technical Data	*283*
9.4.1.1	Identity and Nature of the Source Material	*283*
9.4.1.2	Manufacturing Process	*284*
9.4.1.3	Chemical Composition	*284*
9.4.1.4	Specifications	*284*
9.4.1.5	Stability of the Botanical or Botanical Preparation used as Ingredient in Food Supplement	*284*
9.4.1.6	Proposed Uses and Use Levels	*284*
9.4.1.7	Information on Existing Assessments	*285*
9.4.2	Additionally Data on Exposure and Toxicological Exposure: Extent and Time	*285*
9.4.3	Toxicological Data	*285*
9.5	Conclusions	*286*
	References	*286*

Index *291*

Preface

In December 2000 Dr. W. Johannisbauer (Cognis GmbH, Düsseldorf) started a discussion at GVT (Forschungs-Gesellschaft Verfahrens-Technik e.V.) in respect of the topic "liquid–solid extraction". The issue was to focus activities in an area having wide industrial application, that is, extraction of natural substances, of flavorings, fragrances, APIs (active pharmaceutical ingredients) etc. A series of symposia was held concerning extracts from plant materials, at DECHEMA, Society for Chemical Engineering and Biotechnology, Frankfurt. A report can be retrieved from the homepage of the ProcessNet Working Party "Extraction" (http://www.processnet.org/extraktion). In 2010 the people organizing these activities founded the ProcessNet Working Party "Phytoextrakte – Produkte und Prozesse" (http://www.processnet.org/FDTT.html), where S. Pilz is one of the chairpersons. However, in recent years more oral presentation on that topic could be heard at the annual ProcessNet meetings. The keynote lecture "Phytoextraktion von Naturstoffen" (Extraction of Natural Products) by H.J. Bart and M. Schmidt (both from TU Kaiserslautern) and F. Bischoff (Boehringer Ingelheim) initiated this book project with Wiley-VCH, Weinheim.

The book is for an experienced reader from academic or industry, who has basic knowledge in conventional liquid–liquid extraction. The first chapter is an introduction to extraction and purification methods and aspects of cultivation. The following chapter gives the state-of-the-art on methods to calculate and predict the solubility of complex natural and pharmaceutical products in solvent and solvent mixtures, in supercritical fluids at elevated pressures, as well as at different pH values. The third chapter is on alternative solvents with a special focus on surfactant solutions to achieve benign extraction conditions. Chapter 4 exclusively deals with high pressure processing with supercritical fluids and is followed by process engineering (extraction, chromatography, etc.) and miniplant technology. The technological aspects (modeling, scale-up, industrial equipment . . .) are given in more detail in Chapter 6. This is followed by an actual application example with the extraction of lignocelluloses and algae and followed by a chapter on the market development and potentials. The book closes with regulation and requirements, which are of vital importance in respect to foodstuff and which differ from country to country.

The book, written by experts, is intended to serve as a kind of handbook of industrial (not analytical) scale extraction of natural products. The large number of references, regulations and directives should support the reader and supply the state-of-the art in this field.

We would like to thank all the contributors for their well-written articles and Wiley-VCH for their patience in the production process. We hope that this book will serve to stimulate academic and industry and act as basis and new focus to promote extraction of natural products in the future.

March 2011   *Hans-Jörg Bart[1] and Stephan Pilz[2]*
[1]TU Kaiserslautern and
[2]Evonik Degussa GmbH

List of Contributors

Werner Bäcker
Bayer Technology Services GmbH
BTS-PT-PT-CEM/Gb310
51368 Leverkusen
Germany

Hans-Jörg Bart
Technische Universität Kaiserslautern
Lehrstuhl für Thermische
Verfahrenstechnik
Gottlieb-Daimler-Straße 44
67653 Kaiserslautern
Germany

Friedrich Bischoff
Boehringer Ingelheim Pharma GmbH
Co. and KG
Pharma Chemicals
55216 Ingelheim am Rhein
Germany

Sybille Buchwald-Werner
Vital Solutions GmbH
Hausinger Straße 4-8
40764 Langenfeld
Germany

Dirk Delinski
RWTH Aachen University
AVT – Thermische Verfahrenstechnik
Wüllnerstraße 5
52062 Aachen
Germany

Rudolf Eggers
Technische Universität
Hamburg-Harburg
Thermische Verfahrenstechnik:
Wärme- und Stofftransport
Eißendorfer Straße 38
Hamburg 21073
Germany

Thomas Hahn
University of Kaiserslautern
Institute for Bioprocess
Engineering
Gottlieb-Daimler-Straße 44
67663 Kaiserslautern
Germany

Wilhelm Johannisbauer
E&V Energie und
Verfahrenstechnik
Erich-Kästner-Straße 26
40699 Erkrath
Germany

Volkmar Jordan
Münster University of Applied Sciences
Department of Chemical Engineering
Stegerwaldstraße 39
48565 Steinfurt
Germany

Horst Josten
Cognis GmbH
Henkelstraße 67
40589 Düsseldorf
Germany

Svenja Kelly
University of Kaiserslautern
Institute for Bioprocess Engineering
Gottlieb-Daimler-Straße 44
67663 Kaiserslautern
Germany

Alwine Kraatz
University of Applied Sciences
Department of Nutrition and Home
Economics – Oecotrophologie
Corrensstraße 25
48149 Münster
Germany

Kai Muffler
University of Kaiserslautern
Institute for Bioprocess Engineering
Gottlieb-Daimler-Straße 44
67663 Kaiserslautern
Germany

Ulrich Müller
Hochschule Ostwestfalen–Lippe
Fachbereich Life Science Technologies
Liebigstraße 87
32657 Lemgo
Germany

Andreas Pfennig
RWTH Aachen University
AVT – Thermische Verfahrenstechnik
Wüllnerstraße 5
52062 Aachen
Germany

Stephan Pilz
Evonik Degussa GmbH
Rodenbacher Chaussee 4
63457 Hanau-Wolfgang
Germany

Guido Ritter
University of Applied Sciences
Department of Nutrition and Home
Economics – Oecotrophologie
Corrensstraße 25
48149 Münster
Germany

Feelly Ruether
Technische Universität Dortmund
Laboratory of Thermodynamics
Emil-Figge-Straße 70
44227 Dortmund
Germany

Gabriele Sadowski
Technische Universität Dortmund
Laboratory of Thermodynamics
Emil-Figge-Straße 70
44227 Dortmund
Germany

Michael Schulte
Merck KGaA
PC-R
Frankfurter Straße 250
64293 Darmstadt
Germany

Jochen Strube
TU Clausthal
Institut für Thermische
Verfahrenstechnik
Leibnizstraße 15a
38678 Clausthal-Zellerfeld
Germany

Nils Tippkötter
University of Kaiserslautern
Institute for Bioprocess Engineering
Gottlieb-Daimler-Straße 44
67663 Kaiserslautern
Germany

Roland Ulber
University of Kaiserslautern
Institute for Bioprocess Engineering
Gottlieb-Daimler-Straße 44
67663 Kaiserslautern
Germany

1
Extraction of Natural Products from Plants – An Introduction
Hans-Jörg Bart

1.1
Introduction

The history of the extraction of natural products dates back to Mesopotamian and Egyptian times, where production of perfumes or pharmaceutically-active oils and waxes was a major business. In archeological excavations 250 km south of Baghdad extraction pots (Figure 1.1) from about 3500 BC were found [1], made from a hard, sandy material presumably air-dried brick earth. It is supposed that in the circular channel was the solid feed, which was extracted by a Soxhlet-like procedure with water or oil. The solvent vapors were condensed at the cap, possibly cooled by wet rags. The condensate then did the leaching and was fed back through holes in the channel to the bottom.

Several Sumerian texts also confirm that a sophisticated pharmaceutical and chemical technology existed. In the oldest clay tablets of 2100 BC (Figure 1.2), found 400 km south of Baghdad, is a description of a simple batch extraction: "purify and pulverize the skin of a water snake, pour water over the amashdubkasal plant, the root of myrtle, pulverize alkali, barley and powered pine tree resin, let water (the extract) be decanted; wash it (the ailing organ) with the liquid; rub tree oil upon it, let saki be added" [2].

The pulverization, admixture of NaCl, alkali (potash) or as reported also KNO_3, shows some basic knowledge of chemical engineering [3]. Also well-documented are recipes to obtain creams and perfumes, from the time of the Assyrian king Tukulti-Ninurta I, 1120 BC. The natural feedstock was crushed in a mortar, and then leached in boiled water for one day. New feed was then added gaining higher concentrations. After percolation, oil was added while increasing the temperature. After cooling, the top oil extract can be removed, and the use of demisters (sieves of clay filled with wool or hair) is also reported [4–6].

In a papyrus of 1600 BC, beer and wine were used as alcoholic solvents [7], which give the distinct advantages of achieving a higher solubility for a solute when producing perfumes. Otherwise up to 40 repetitive extraction procedures as above were necessary to give a high yield. There was not much development until medieval times, when pure ethanol became available as a solvent in about 900 AD [3].

Industrial Scale Natural Products Extraction, First Edition. Edited by Hans-Jörg Bart, Stephan Pilz.
© 2011 Wiley-VCH Verlag GmbH & Co. KGaA. Published 2011 by Wiley-VCH Verlag GmbH & Co. KGaA.

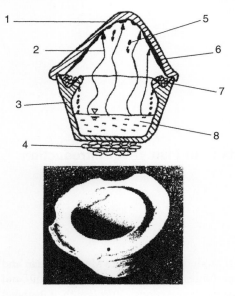

Figure 1.1 Extraction pot (oil or water).

Figure 1.2 Sumerian text (2100 BC).

After this short historical review we should consider the language used. In German "natural plant extraction" is equivalent to "phyto-extraction", which in English terms means extraction (of e.g., metal ions from soil) achieved by plants. The term "natural products" is perhaps also something of a misnomer [8]. A natural product is a chemical compound or substance produced by a living organism. They may be extracted from tissues of terrestrial plants, marine organism or microorganism fermentation [9]. In that respect any biological molecule is a natural product, but in general the term is reserved for secondary metabolites (carotinoids, phytosterines, saponines, phenolic compounds, alkaloids, glycosinates, terpenes etc.), produced by an organism. They are not essential for normal growth, development or reproduction and its survival.

Besides venoms, toxins, and antibiotic peptides from animals (frogs, spider, snake etc.), a new focus is nowadays on the marine world (e.g., curacin A with antitumor activity from marine cyano bacteria). The use of micro-organisms is already well-established in industrial fermentation [10]. Alternative to this, extracts from plant tissue are a rich source of lead compounds for nutraceutical or pharmaceutical applications [11, 12]. The market alone for herbs for nutritional supplements, for example, green tea, melissa, blueberry, is about 6.7 billion euros in Europe and 17.5 billion worldwide. In respect to this, the average trading volume of medicinal plant raw material is, according to FAO, 1 billion US$, and was 440 million US$ in USA in 1997. The annual growth rates for nutraceuticals and pharmaceuticals derived by industrial product extraction is about 6 to 8% (see Table 1.1) [13]. Here triterpenes have the highest growth rates (Figure 1.3). In 2002 the world market for pharmaceuticals from natural plants was estimated to be 30.7 billion US$, and the share of triterpenes is given in Figure 1.4 and the market in Europe depicted in Figure 1.5 [14].

The feed material for the extraction of natural products from plants can be leaves, flowers, branches, bark, rhizomes, roots, seed and fruits, and the active pharmaceutical ingredient (API) content is usually in the region from 0.3 to 3% with

Table 1.1 Market and annual market growth.

Year/annual growth	1993	1998	2003	2008	1993–1998	1998–2002
Chemicals from natural plants [1000 tons]	168	243	336	451	7.7%	6.7%
Average price [US$ kg^{-1}]	7.05	7.78	8.74	9.96	2.0%	2.3%
Demand for natural products [10^6 US$]	1185	1890	2935	4495	9.8%	9.2%
Essential oils [10^6 US$]	465	625	820	1054	6.1%	5.6%
Extract from plants [10^6 US$]	268	560	1120	1990	15.9%	14.9%
Gums, gels, polymers [10^6 US$]	284	392	500	660	7.4%	5.0%
Others [10^6 US$]	178	313	495	791	11.9%	9.6%

1 Extraction of Natural Products from Plants – An Introduction

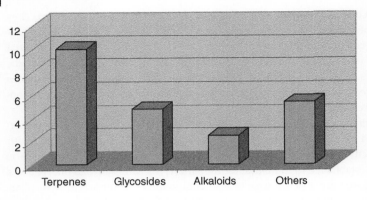

Figure 1.3 Annual growth rates (1997–2002).

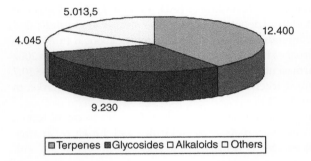

Figure 1.4 Market share of active pharmaceutical ingredients.

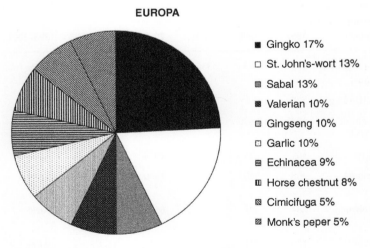

Figure 1.5 Natural plant extracts in Europe (2002).

Table 1.2 API in plants.

Leaves	Bark	Roots	Seed/fruits
Belladonna	Chinchona	Licorice	Senna
Hyoscyamus		Rauwolfia	Oenothera
Duboisia		Ipecac	Horse chestnut
Digitalis		Berberis	Sabal
Senna		Ginseng	
Catharanthus		Valerian	
Ginkgo			
Echinacea purpurea			
Cimifuga			
Hawthorn			

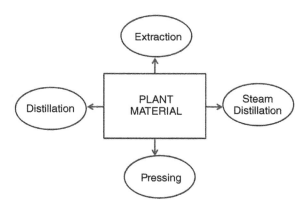

Figure 1.6 Methods for recovery of secondary metabolites.

seasonal fluctuations in period and producing area. Table 1.2 gives an overview, where the API can be found in the plant materials. The recovery of secondary metabolites is depicted in Figure 1.6. The most economic route is via (cold) pressing, mainly for oily constituents. Volatile compounds can be removed either by conventional (vacuum) distillation or hydrodistillation. The vapors from steam distillation are condensed and a two-phase distillate (oil and water) is obtained. The boiling point of this practically immiscible oil–water mixture is always below 100 °C, since each constituent independently exerts its own vapor pressure as if the other constituent is not present and thus high boiling thermosensitive oils (e.g., lavender oil and other fragrances) can be efficiently prepared. In the following we will concentrate on solids extraction in order to generate a primary crude extract.

1.2 Cultivation

The production chain with natural products has three major parts to consider. The first is agricultural, followed by, for example, extraction to get a concentrated raw extract, and in pharmaceutical applications (not with cosmetics and nutraceuticals) a final purification step is necessary in order to get an ultrapure product. All the steps combined contribute to the overall yield and determine the final economics.

To highlight the importance of including the complete production chain, we can consider growing a crop of plants containing oleanoic acid. This is found in several plants (almond hulls, privet, rosemary, thyme, clove, lavender, olive, hawthorn, periwinkle etc.) with a content less than 1% in dry mass. Higher values were observed only with a different kind of sage. The value varies with location and local climate (e.g., Germany versus Greece), types of soil (sandy, thus irrigation needed), distance between the growing rows and cultivar. With the latter, higher content is with plants from scions obtained through vegetative reproduction (*Salvia officinalis*) compared with those from sowing (*S. lavendulifolia*). However, planting with scions is very labor-intensive, which has quite an impact in respect to labor costs, and sowing is then the alternative. Optimal conditions for sage are with warm and wind-protected sites, with a light soil containing compost and water. Dry periods are no major problem and the plants should grow in rows (distance 25–50 cm) as depicted in Figure 1.7 [15]. Usually, the useful life is four to five years and for soil recovery a four-year interval is recommended. If the plants are used to gain pharmaceutical extracts, all cultivation steps (sowing, manuring etc.) must be documented and with pest management any pesticide (date, dosage etc.) must be registered. With sage the use of herbicides is forbidden and two cuts per year are recommended. After air classification the leaves are dried and according to the *Deutsches Arzneibuch* only 2% impurities are

Figure 1.7 Cultivation tests of sage at DLR Rheinpfalz.

allowed [16]. A typical yield in an eight-year test cultivation in Saxony is given in Table 1.3 [17].

In a three-year project, TU Kaiserslautern, DLR (Dienstleistungszentrum Ländlicher Raum) Rheinpfalz, and the local pharmaceutical industry investigated the cultivation and extraction of sage with respect to the recovery of APIs, like ursolic and oleanoic acid. They have a similar structure, as they are position isomers, since only a CH_3-group is shifted (Figure 1.8). This makes a final separation very difficult in order to obtain ultrapure products.

The general trends found in Saxony for the harvested plants could be confirmed. Interestingly, the overall yield after the first year remained almost constant with all plants, but there is a strong dependency on the triterpene content with harvest time. The first cut in June contains mainly monoterpenes and can be further processed for tea, spice etc. The second cut in September shows an increased triterpene content (maximum here for both acids is about 6%) and even better results could be achieved if the plant is under a poly-tunnel (because of the higher local temperature). In Figure 1.9 triterpene content for *S. officinalis* and *S. lavendulifolia* is compared [14].

Table 1.3 Quality and yield in *Salvia officinalis* under the climate in Saxony.

Fresh leaves[a]	1. year: 4–12 t/ha
	2. year: 8–24 t/ha
Dried leaves[a]	1. year: 1–3 t/ha
	2. year: 2–4.5 t/ha
Essential oil	1.2–2.5%
α-thujone	25–42% in essence
Camphor	12–21% in essence
Rosmarine acid	0.4–3.4%
Carnosine derivates	2.3–3.4%
Flavanoides	0.5–1.1%

a) second cut gave 1/3 yield of the first cut.

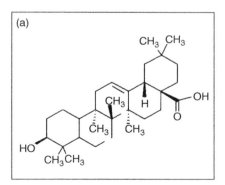

Figure 1.8 Ursolic (a) and oleanoic acid (b).

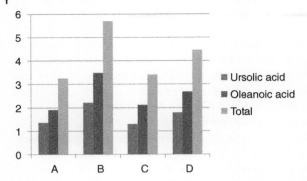

Figure 1.9 Triterpene content in *Salvia officinalis* (A, B) and *S. lavendulifolia* (C,D) after the first (A,C) and second cut (B,D).

1.3
Extraction

1.3.1
Solvents

After harvesting the next step in the process chain is the extraction of the desired substance. However, there are several regulations to be considered (see also Chapter 9). If the products are to be used with foods then there are regulations in the use of appropriate solvents. According to European Union and governmental regulations [18] the following solvents are allowed:

- water (with admixture of acids or base),
- other foodstuffs with solvent properties and
- solvents like propene, butane, ethylacetate, ethanol, CO_2, N_2O, acetone (the latter not with olive oil).

With water and foodstuffs it is assumed that the inevitable solvent residue is not harmful. In contrast to this, with industrial solvents maximum residues are defined for a certain purpose, as given in Table 1.4. In Table 1.5 one finds the limits when aroma is extracted from natural resources. In any case, for all these solvents the maximum content of arsenic or lead is 1 mg and no toxicologically critical additives are allowed. The use of a mixture of hexane and ethylmethyl ketone is forbidden.

As can be seen from above water, solvents from natural sources (limonene etc.), organic solvents and liquefied gases are used in the food industry. Here liquefied CO_2 dominates the market and is used for decaffeination of green coffee beans or tea, preparation of leaf extracts, extraction of spices, herbs, essential oils, pungent constituents, natural colorants and antioxidants as well as production of high-value fatty oils [19] (see also Chapter 4 in this book).

Table 1.4 Solvents with foodstuffs and maximal residue content.

Solvent	Purpose	Max. residue
Hexane	Fractionating of fats, oils or cacao butter	1 mg kg^{-1} in oil, fat or cacao butter
	Defatting of protein containing products respectively flour	30 mg kg^{-1} in defatted soy products, otherwise 10 mg kg^{-1}
	Defatting of corn seed	5 mg kg^{-1} in defatted seed
Methylacetate	Extraction of for example, caffeine or other bitter constituents from tea or coffee	20 mg kg^{-1} in coffee or tea
	Production of sugar from molasses	1 mg kg^{-1} sugar
Ethylmethylketone	Fractionating of oils and fats	5 mg kg^{-1} in oil or fat
	Extraction of for example, caffeine or other bitter constituents from tea or coffee	20 mg kg^{-1} in tea or coffee
Dichloromethane	Extraction of for example, caffeine or other bitter constituents from tea and coffee	2 mg kg^{-1} in roasted coffee and 5 mg kg^{-1} in tea
Methanol	For all products	10 mg kg^{-1}
Propane-2-ol	For all products	10 mg kg^{-1}

Table 1.5 Residues in artificial flavored products.

Solvent	Max. residue (mg kg^{-1})
Diethylether	2
Hexane	1
Cyclohexane	1
Methylacetate	1
Butane-1-ol	1
Butane-2-ol	1
Ethylmethylketone	1
Dichloromethane	0.02
Propane-1-ol	1
1,1,1,2-Tetrafluoroethane	0.02

Similar guidelines, which recommend acceptable amounts for residual solvents in pharmaceuticals also exist as distributed by the European Medicines Agency [20]. Solvents that are known to cause unacceptable toxicities (class 1, e.g., benzene, carbon tetrachloride, 1,2-dichloroethane, 1,1-dichloroethane, 1,1,1-trichloroethane)

Table 1.6 Class 2 solvents in pharmaceutical products.

Solvent	PDE (mg/day)
Acetonitrile	4.1
Chlorobenzene	3.6
Chloroform	0.6
Cyclohexane	38.8
1,2-Dichloroethene	18.7
Dichloromethane	6.0
1,2-Dimethoxyethane	1.0
N,N-Dimethylacetamide	10.9
N,N-Dimethylformamide	8.8
1,4-Dioxane	3.8
2-Ethoxyethanol	1.6
Ethylene glycol	6.2
Formamide	2.2
Hexane	2.9
Methanol	30.0
2-Methoxyethanol	0.5
Methylbutylketone	0.5
Methylcyclohexane	11.8
N-Methylpyrrolidone	48.4
Nitromethane	0.5
Pyridine	2.0
Sulfolane	1.6
Tetraline	1.0
Toluene	8.9
1,1,2-Trichloroethene	0.8
Xylene[a]	21.7

a) usually 60% m-xylene, 14% p-xylene, 9% o-xylene with 17% ethyl benzene.

should be avoided in the production of drug substances, excipients, or drug products, unless their use can be strongly justified in a risk–benefit assessment. Some solvents associated with less severe toxicity (class 2) should be limited in their usage. Here Table 1.6 gives the "permitted daily exposure" (PDE), which is a guideline for the pharmaceutically acceptable intake of residual solvents in mg/day. Ideally, less toxic solvents (class 3) should be used where practical (Table 1.7). However, there remain solvents, which may also be of interest to manufacturers of excipients, drug substances, or drug products, as given in Table 1.8. However, no adequate toxicological data on which to base a PDE was found. Manufacturers should supply justification for residual levels of these solvents in pharmaceutical products.

However, more exotic solvents, like surfactant rich phases or ionic liquids are applied to the extraction of plants (see Chapter 3). The latter is a new class of solvents with almost no vapor pressure, and surfactant rich phases enable the

Table 1.7 Class 3 solvents which should be limited by GMP or other quality based requirements.

Acetic acid	Heptane
Acetone	Isobutyl acetate
Anisole	Isoproyl acetate
1-Butanol	Methyl acetate
2-Butanol	3-Methyl-1-butanol
Butyl acetate	Methylethyl ketone
Tert-Butylmethyl ether	Methylisobutyl ketone
Cumene	2-Methyl-1-propanol
Dimethylsulfoxide	Pentane
Ethanol	1-Pentanol
Ethyl acetate	1-Propanol
Ethyl ether	2-Propanol
Ethyl formate	Propyl acetate
Formic acid	Tetryhydrofuran

Table 1.8 Solvents for which no adequate toxicological data exist.

1,1-Diethoxypropane	Methylisopropylketone
1,1-Dimethoxymethane	Methyltetrahydrofuran
2,2-Dimethoxypropane	Petroleum ether
Isooctane	Trichloroacetic acid
Isopropyl ether	Trifluoroacetic acid

extraction of large biomolecules (proteins etc.), which like an aqueous environment and would degenerate in any other solvent. However, the use of water is ambiguous. Traditional medicines are often prepared by water extraction, but water-soluble impurities present challenges for conventional isolation methods, such as chromatography or crystallization. Water preferentially extracts polar compounds (e.g., plant pigments, tannins) and they need some special post treatment (e.g., ion exchange, caustic wash [21]) for further purification. In many cases the crude extract is evaporated and the residue is dissolved in an appropriate solvent for further purification. Here water has the highest heat of evaporation and is more difficult to remove compared with an organic solvent.

The choice of the solvent often follows the principle "like extracts like". According to this concept one will find the concept of solubility parameters, based on the regular solution theory, first developed by van Laar [22] to have an indication for choosing any solvent. However, the results showed strong dependence on the mixing rules applied and was improved by Hildebrand and Scott [23]. They replaced the two van der Waals parameters with two new ones, the molar volume V_i and the solubility parameter δ_i of a solute i.

The solubility parameter δ_i is defined as:

$$\delta_i = \sqrt{\frac{\Delta U_i}{V_i}} \tag{1.1}$$

where ΔU_i is the internal energy of vaporization and at normal pressures related to the heat of vaporization, ΔH_{evap}, as follows:

$$\Delta U_i \approx \Delta H_{evap} - RT \tag{1.2}$$

For multicomponent mixtures the Hildebrand and Scott model is:

$$RT \ln \gamma_i = V_i \left(\delta_i - \bar{\delta}\right)^2 \tag{1.3}$$

with the mean solubility parameter of the solvent:

$$\bar{\delta} = \sum_{k=1}^{n} \varphi_k \delta_k \tag{1.4}$$

where φ is the volume fraction of the component k in the mixture which can be calculated from the molar volumes, V_i, of all solutes and the solvent. A Flory–Huggins correction considers the effects of different molar volumes:

$$\ln \gamma_i = \frac{V_i}{RT}\left(\delta_i - \bar{\delta}\right)^2 + \ln\left(\frac{V_i}{V_m}\right) + 1 - \frac{V_i}{V_m} \tag{1.5}$$

with

$$V_m = \sum_{k=1}^{n} x_k V_k \tag{1.6}$$

These effects exceed 10% if the molar volumes of the solutes differ more than 50% with respect to the solvent, V_m and otherwise the use of this correction is recommended. The regular solution model is suitable only for non-polar mixtures of molecules that are not too different in size, the activity coefficients always exceed unity, and the excess enthalpy is always positive. Nevertheless, even in the absence of any other information the predictive model is quite useful in certain cases. A compilation of solubility parameters is given by Barton [24]. However, literature provides quite a number of solubility parameters [25, 26], developed in the polymer field etc. However, this concept certainly has a limited application with natural products, which are usually polar, have functional groups and the ability to form hydrogen bonds. Here the concept based on solvatochromic scales is more promising.

At a molecular scale a solvation process will take place in several stages, although only the overall process is measurable. First, a cavity must be created in the solvent to accommodate the solute. Donor–acceptor bounds between the solute and solvent will develop and dipole orientations will be induced in non-polar but polarizable solvent molecules and dipolar solutes. Once this new aggregate is formed the solvated solute may further interact with its surrounding (hydrogen bonding etc.). The Gibbs free energy change for the process of dissolution of a solute i from phase α to β is zero at equilibrium:

$$0 = \Delta_{\alpha,\beta}\mu_{s,i} = \frac{\mu_i^{\infty,\alpha} - \mu_i^{\infty,\beta}}{RT} + \ln\frac{x_i^\alpha}{x_i^\beta} + \ln\frac{\gamma_i^{\infty,\alpha}}{\gamma_i^{\infty,\beta}} \tag{1.7}$$

At infinite dilution the activity coefficients approach unity and with the definition of the distribution coefficient:

$$K_i^{\alpha/\beta} = \frac{x_i^\alpha}{x_i^\beta}\frac{\gamma_i^\alpha}{\gamma_i^\beta} \tag{1.8}$$

and

$$\lim_{x \to 0} \ln K_i^{\alpha/\beta} = \ln K_i^{\infty,\alpha,\beta} = \frac{\mu_i^{\infty,\beta} - \mu_i^{\infty,\alpha}}{RT} \tag{1.9}$$

The standard molar Gibbs free energy of solvation, $\Delta\mu_{s,i}^\infty$, can be derived from pure component data using spectroscopic information for determining solvatochromic parameters. A generalized equation for $\Delta\mu_{s,i}^\infty$ with a linear dependence on solvatochromic parameters is:

$$\Delta\mu_{s,i}^\infty = C_0 + C_1\delta_i^2 + C_2\pi_i^* + C_3\alpha_i + C_4\beta_i \tag{1.10}$$

Here δ_i^2 is the cohesive energy density of the solvent which is the square of the Hildebrand solubility parameter; α_i and β_i characterize respectively activity and basicity which in general represents the ability to form hydrogen bonds; π_i^* defines the polarity or polarizability of the solvent.

The estimation of solvatochromic parameters is based on the empirical rule that absorption signals in spectrometry are shifted due to solvent interactions. The estimation of solvatochromic parameters is on an empirical basis which says that the position, ξ, of different solutes (indicators) in spectra of a solvent is a linear function:

$$\xi = C_0 + C_1\pi_i^* + C_2\alpha_i + C_3\beta_i \tag{1.11}$$

$C_0, \ldots, C_3$ are properties of the indicator and π^*, α_i and β_i are properties of the solvent. With different analytical methods (NMR, UV-VIS etc.) and different indicators the solvatochromic parameters of a solvent can be estimated. These linear relationships not only correlate spectroscopic positions of indicators in different solvents, but can also be used to correlate the influence of the solvent on reaction equilibria and solubilities. In principle it considers differences in the Gibbs free energy resulting from a change of the electron configurations in a molecule.

There exists a number of such linear solvatochromic scales. One of the most widely used is that of Kamlet and Taft which is the basis of the LSER (linear solvation energy relationships) [27, 28]. The Nernst distribution (concentration ratio of the solute in organic to aqueous phase) according to Kamlet is:

$$\log D_i^\infty = C_0 + C_1 V_i + C_2 \partial_i + C_3\pi^* + C_2\alpha_i + C_2\beta_i \tag{1.12}$$

Here the cohesive energy density in Equation 1.10 is replaced by the molar volume V_i of the solute (as a measure of the size of the cavity to accommodate the solute i in the solvent) and ∂_i is an empirical parameter which takes also account for polarizability π^*. A major disadvantage of the methods above is the limited predictivity and applicability of those correlations, since they have not been developed in the field of natural substances. However, a newer approach relies on the equation of state concept as is discussed in Chapter 2, with promising results in respect to predictability of solubilities.

Besides solubility, which is a key feature in obtaining a crude extract with any solvent, there are additional criteria for solvent selection, similar to those in liquid–liquid extraction.

- **Selectivity** – high selectivity enables fewer stages to be used. If the feed is a complex mixture where multiple components need to be extracted, group selectivities become important.
- **Recoverability of solvent** – recovery of the solvent phase should be easy. If evaporation or distillation is used the solvent should have a low heat of evaporation, should not form azeotropes, and be easily condensed by cooling water. With liquid CO_2 or other subcritical fluids only a flash is necessary. Ionic liquids cannot be evaporated thus the product must be volatile. However, if alkylated tertiary amines (e.g., trioctylamine) is used in acidic media (e.g., HCl), the ionic liquid forms by itself (e.g., trioctylammoniumchloride). This reaction can be reversed by alkali and the free amine can then be distilled. If surfactant phases are used, a recovery is then with ultrafiltration.
- **Viscosity and melting point** – high viscosities reduce the mass transfer efficiency and lead to difficulties with pumping and dispersion. The melting temperature of the solvent should preferably be lower than ambient for ease of handling.
- **Surface tension** – low surface tension promotes wetting of the solids. Wetting ability is important since the solvent must penetrate the matrix (pores, capillaries etc.).
- **Toxicity and flammability** – for food processing only nontoxic solvents will be taken into consideration. In general, any hazard associated with the solvent will require extra safety measures. As to this, aliphatic diluents are preferred to aromatic or halogenated ones.
- **Corrosivity** – corrosive solvents increase equipment cost but might also require expensive pre- and post-treatment of streams.
- **Thermal and chemical stability** – it is important that the solvent should be thermally and chemically stable as it is recycled. Especially it should resist breakdown during the solvent recovery in an evaporator.
- **Availability and costs** – solvent should be ready available. It is not the price of the solvent that is important, but the annual cost due to the inevitable operation losses.

- **Environmental impact** – the solvent should not only be compatible with downstream process steps, but also with the environment (minimal losses due to evaporation, solubility and entrainment). Removal of solvents from residual plant material (and spent solids) can cause serious problems and post treatment may be necessary to reduce the residue level. This is mainly by mechanical pressing since further treatment using another solvent is not economic.

1.4
Extraction Techniques

Leaching or solid–liquid extraction is a separation process where often no sharp interface for mass transfer can be defined. In many natural product extraction processes the "solid" contains or is impregnated by a liquid phase. In the extraction of sucrose from beet the cell walls prevent undesired and high molecular constituents from being extracted. Thus the beet is prepared in long strips in order to minimize damage to the cell walls. If fatty oils are extracted the solute is itself a liquid and will diffuse more easily resulting in a faster mass transfer. The latter is markedly affected if the solute is solid or liquid, incorporated within, chemically combined with, adsorbed upon, or mechanically in a pore structure of the insoluble material [29].

In that respect a pre-treatment to change the initial matrix structure is generally recommended. The classical approach is with grinding (bead milling) to gain smaller particles, which can be more easily penetrated by the solvent. Alternatively, enzymatic treatments, freezing and thawing or very often a swelling process in pure water gives an improvement. In Figure 1.10 ground leaves of *S. lavendulifolia* were soaked at pH = 7, 3 and 12 and the maximum yield in an alcohol extract was with pure water after four hours [30]. However, soaking of fresh plant material with organic solvents (ethanol, methanol) is recommended, since enzymes will be denatured, preserving the solute undamaged. More drastic approaches use hot

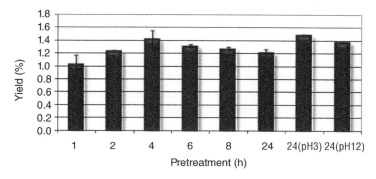

Figure 1.10 Pre-treatment at 298 K (pH = 7, pH = 3 and pH = 12).

steam or cooking. Materials destined for oil production containing large amounts of proteins (e.g., cottonseeds, soya beans, flax seeds, sesame seeds, peanuts) must be cooked in order to coagulate the proteins before oil extraction. Expanding vapors from any pressurized apparatus increase porosity due to explosion, like evaporation from the solid cells [31]. An alternative to this is to use pulsed electric fields [32] or ultrasound [33] to destroy the cell walls and facilitate extraction. However, this is mainly used in bench-scale equipment [34], since with such fields it is difficult to adjust penetration depth and energy efficiency at larger scales.

In terms of extraction procedures one has to distinguish between batch and continuous operations. The latter are frequently applied with suspensions or slurries but apparatuses for continuous solid extraction are also widely used [31, 35].

Maceration is preferably used with volatile or thermal instable products, it is a "cold" extraction of pulverized feed material in any solvent. A well-known example is the extraction of color, tannin and aroma from red grape skins by alcohol during the fermentation process, or is used to produce any perfume stock. This can be supported by enzymatic processes, as discussed in Chapter 7 in this book. If this process is then at higher temperatures it is called digestion, which in daily life everybody is familiar with tea preparation. The main disadvantage of this process is the sometimes long duration (hours) and the solvent consumption in a batch stirred tank.

When the mean diameter of the feed material increases percolation is a widely used technique. The solvent is poured on top of the solid material and allowed to percolate through the bed. However, fine powders and plants may swell (e.g., containing mucilages) and can clog the percolator. The apparatus side (see also Chapter 6 in this book) is well established with carrousels, baskets, sliding cells etc. and found in daily life with for example, filter coffee preparation.

As mentioned at the beginning, a very efficient form is Soxhlet extraction [36] which was originally designed for the extraction of a lipid from solid material. The solvent is heated to reflux (condensed) and percolates the solid material. The disadvantage is that the solute is always at the boiling temperature of the solvent, which may cause damage of thermolabile compounds. An alternative to this is distillation, which can only be used with volatile compounds. The vapors from steam distillation are condensed and a two-phase distillate (oil and water) is obtained. The boiling point of this practically immiscible oil–water mixture is always beneath 100 °C, since each constituent independently exerts its own vapor pressure as if the other constituent is not present and thus high boiling thermosensitive oils (e.g., lavender oil and other fragrances) can be efficiently prepared.

The mathematical treatment of mass transfer and co- and counter-current apparatus balances is very often found also under "leaching" [35]. It is sometimes convenient to use right-angled triangular diagrams to represent the process. In Figure 1.11 a counter-current process is depicted. In a multistage operation the overall apparatus balance is

$$F + E_i = R_i + E_1 \tag{1.13}$$

or

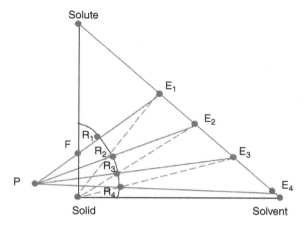

Figure 1.11 Right-angled triangular diagram for counter-current solid extraction.

$$P = F - E_1 = R_i - E_i \qquad (1.14)$$

Where F is the feed, E_1 is the loaded and E_i the fresh solvent and R_i is the raffinate after the ith stage. All the R_i lie on a pseudo-equilibrium curve, which has to be determined experimentally. The true equilibria with solids extraction are often difficult to determine, due to the sometimes very slow extraction kinetics (see also Chapter 6 in this book). The construction of the stages is similar to the Hunter–Nash concept with liquid–liquid extraction [37]. However, since the solvent is not soluble in the solid all the tie lines meet at the solid corner. This is depicted in Figure 1.11 with four stages of extraction.

1.5
Purification

The purification methods relay mainly on chromatography and the final product is then obtained by crystallization (see also Chapter 5 in this book). In applications with nutraceuticals, cosmetics and fragrances there is no needs for ultrapure products, which is in strong contrast to the pharmaceutical field. Chromatographic methods are very flexible due to their separation principles, which are described below.

1.5.1
Chromatography

1.5.1.1 Adsorption Chromatography
Adsorption chromatography or liquid–solid chromatography is the oldest form of chromatography. The sample components are adsorbed on the surface of the

adsorbent and displace the initially loosely adsorbed solvent molecules. Adsorption chromatography gives best results for non-ionic, organic-soluble samples of intermediate molecular weights. Water-soluble substances can also be separated satisfactory, but one of the other methods gives usually better results. The stationary phase (the adsorbent) is often a polar solid (almost always silica or alumina), that is used in connection with a non-polar liquid. The (polar) functional groups of the sample components are then easily attracted by the adsorbent surface and displace the non-polar solute molecules. The quality of the separation strongly depends on the solute used. For low pressure separations also charcoal is used. Here we have a non-polar adsorbent that is used with a non-polar mobile phase. The adsorption depends on the polarizability of the sample molecule and thus compounds of higher refractive index are stronger adsorbed.

1.5.1.2 Partition Chromatography

Partition chromatography can be divided into *liquid–liquid chromatography (LLC)* and *chromatography on chemically bonded phases*. It involves two immiscible liquids, one as the mobile phase and the other being fixed on a solid support as the stationary phase. The principle of the separation is an extraction process. The stationary liquid can either be fixed on the solid by physical adsorption (LLC), which is easy, but leads to a continuous loss of stationary phase, or by chemical bonds, which is now the most often applied method.

In the case of a polar stationary phase (usually on a silica or alumina support) and a non-polar mobile phase, the process is referred to as *normal phase (NP)* (liquid) liquid chromatography, for historical reasons. In the opposite case it is called *reversed phase (RP)* (liquid) liquid chromatography. Partition chromatography is suitable for a great number of substances in a wide polarity range.

Two special cases of partition chromatography are *chiral liquid chromatography*, which has gained great economic importance for separating racemates in drug generation, and *ion pair chromatography*. Ion pair chromatography is a kind of RP chromatography, that is used for the separation of ionic substances. Here the ions to be analyzed or separated are retained on the stationary phase together with their counter-ions that are delivered with the mobile phase. Desorption is often done with an aqueous solution containing a water-soluble organic solvent like methanol. *Ion pair chromatography* often competes with *ion exchange chromatography*.

1.5.1.3 Ion Exchange Chromatography

Here the two-phase system is made by putting swollen particles of an ion exchanger in contact with an aqueous solution of a mixture of components. If the components form ions in the solution, then electrostatic interactions take place with the ionogenic functional groups of the ion exchanger, which is accompanied by ion exchange. The strength of the interaction is influenced by the charge of the ionogenic component, that is, by the charge on the ion and the dissociation constant. Further on, the process is diffusion limited and thus influenced by the diffusion coefficient of the ions and therefore by the effective size of the hydrated ions. The various types of ion exchangers can be classified as follows [38, 39]:

- cation exchangers
- anion exchangers
- amphoteric and dipolar ion exchangers
- chelating ion exchangers
- selective (or specific) ion exchangers

Cation and *anion exchangers* are the simplest ion exchange resins. They have cations (or anions respectively) bound to a matrix, that can be exchanged with ionic sample components. *Amphoteric ion exchangers* contain both cation and anion exchanging groups in their matrix. These ion exchangers are capable of forming internal salts which dissociate in contact with the sample electrolytes and thus can bind both anionic and cationic components. Dipolar ion exchangers are a special kind of *amphoteric ion exchangers*. In this case, amino acids are bound to the matrix. They form dipoles in an aqueous solution, which interact especially selective with biopolymers. *Chelating ion exchangers* carry functional groups capable of forming a complex bond with metal ions. They bind heavy metals and alkaline earth metals preferentially. *Selective ion exchangers* have a limited binding ability and bind some ions only. *Specific ion exchangers* have a stricter limitation. They react with one type of ion only.

1.5.1.4 Gel Chromatography

Gel chromatography is often called *size exclusion chromatography (SEC)*, because here mixtures are separated due to their unequal size. The stationary phase consists of swollen gel particles with a pore size that enables the small molecules of a sample to penetrate into the pores, while medium size molecules only partially enter the pores and the greatest molecules are completely excluded from the pores. *Gel chromatography* can be used with a hydrophobic mobile phase, it is then also called *gel permeation*, and with a hydrophilic mobile phase, then referred to as *gel filtration*.

1.5.1.5 (Bio-) Affinity Chromatography

Affinity chromatography is the newest variant of liquid chromatography. This technique exploits the unique biological specificity of the protein–ligand interaction. This concept is realized by binding the ligand to an insoluble support just as in *partition chromatography on chemically bonded phases*. The ligands may have low or high molecular weight and consist of nucleic acids, enzymes or many other compounds. Using the prepared substance as the column packing, certain proteins with an appreciable affinity for the ligand will be retained. These can be eluted by altering the composition or pH of the mobile phase to favor dissociation and weaken the binding of the ligand.

1.5.2 Continuous Techniques

The mechanical techniques are with batch, semi-continuous and continuous chromatographic apparatuses [40]. Since batch chromatography is well known and discussed elsewhere [41], a short review is given on continuous concepts.

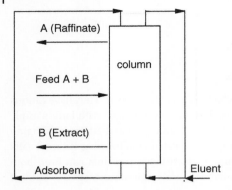

Figure 1.12 True moving bed chromatography.

1.5.2.1 True Moving Bed (TMB) Chromatography

This was the first attempt to achieve a continuous counter-current chromatographic process. The solid phase goes down the column due to gravity; when it exits the system, it no longer contains adsorbed products and is thus recycled to the top of the system. The liquid stream follows exactly the opposite direction. It moves upwards and is recycled from the top to the bottom of the column. Feed, containing the components A and B in the binary case, is injected in the middle and fresh desorbent at the bottom.

Provided that the affinity of A and B for the adsorbent is different (A being the more retained than B), it is possible to adjust the flow rates in order to make B move upward with the fluid stream and A move down with the solid, leading to a spatial separation. This system requires one inlet and two outlet lines (one for the raffinate containing B and one for the extract with purified A). Figure 1.12 shows the principle of the TMB-process.

However, moving bed systems tend to suffer from:

- difficulties in achieving packing flow control
- low mass transfer efficiencies due to an uneven column packing
- packing attrition due to the increased shear forces
- relatively low mobile phase velocities to prevent fluidization of the chromatographic bed.

Because of all these disadvantages, the TMB-process has never been successfully applied. Nevertheless the underlying idea leaded to the development of the widely used simulated moving bed (SMB) process. Knowledge of the TMB process helps in understanding and operating an SMB plant.

1.5.2.2 Simulated Moving Bed (SMB) Chromatography

As will be shown, most of the benefits of a true counter-current operation can be achieved by using several fixed bed columns in series and appropriate shift of the injection and collection points. This is the SMB concept.

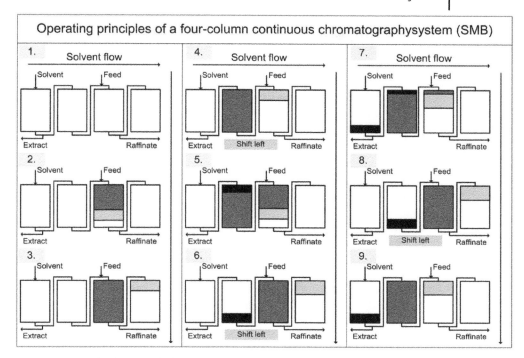

Figure 1.13 Simulated moving bed (SMB) chromatography.

In an SMB plant the inlet and outlet lines are moved step by step between a given number of fixed-bed columns. The solid is no longer moving. Its flow is only simulated by shifting the external lines. In fact, this simulated solid flow rate is directly linked to the shift period. Due to the shifting of the inlet/outlet ports in direction of the liquid flow the solid is still moving with respect to the external lines.

After the start-up period a quasi stationary state is reached, where the concentration profiles move periodically with respect to the inlets and outlets. For an infinite number of columns and an infinitely low shifting time, the profiles are the same is in the TMB process. Figure 1.13 shows the principle of the SMB process.

The feed is pumped continuously into the four-column continuous system, as well as the mobile phase, while two streams of products, the raffinate and the extract, are recovered continuously. The components entering with the feed must exit with one of the products. By focusing on the view from the feed port it becomes apparent that one fraction moves up-stream from the feed to the extract port and the other moves downstream to the raffinate port. In order to keep the concentration profiles within these columns it is necessary to switch the feed and outlet ports all together to the left to get a quasi-stationary operation. The triangular theory of Morbidelli allows to calculate the flow ratios to obtain pure components and a stable operation [42].

The separation operation is carried out at high concentrations. This allows the full use of the non-linear effects, and especially the displacement effect, to enhance separation and production rate. Due to the counter-current operation, the mass transfer between liquid and solid phase is enhanced leading to a better use of the adsorbent and finally to a higher productivity compared to conventional batch processes. Furthermore, the desorbent consumption can be drastically reduced. All these advantages made this process the most widely used among all continuous chromatographic processes. However, the following drawbacks should be recognized:

- high complexity leading to difficulties in operation and making the design difficult and time consuming;
- only two fractions can be collected with a single plant;
- long start-up period.

1.5.2.3 Annular Chromatography

Here the adsorbent is filled between two concentric cylinders which are slowly rotated along their vertical axis. The feed is introduced at a small stationary region at the top of the device. The eluent is also introduced at the top, but is uniformly distributed all along the circumference except the region where the feed is introduced [43].

The rotation affects the separated components to appear as helical bands each of which has a characteristic, stationary exit point. As long as conditions remain constant, the retention time of each component and thus the angular displacement from the fixed feed entry will also remain constant, so that the separated components can be collected at fixed stationary outlets. Hence the separation process is truly continuous. Figure 1.14 shows the principle of this process.

Also the inverse principle, where not the annulus filed with the stationary phase is rotated, but the feed inlet and the outlet ports. The principle is very similar to the classical batch chromatography in a single column. The time dependent separation of batch chromatography is transformed in a spatial angle dependent separation. The productivity of both processes is comparable. The modes of batch chromatography, that is, gradient elution, recycle chromatography, displacement chromatography and frontal analysis can also be applied in annular chromatography. One main advantage of this process in comparison to other continuous chromatographic processes is the simple feasibility of multicomponent separations [44, 45].

1.5.2.4 Carrousel Adsorbers

Several fixed columns are located in a circle on a carrousel. The origin is with ionic separations (ISEP) and liquid adsorption or chromatographic separations (CSEP) [46]. The carrousel with up to 30 columns rotates stepwise, while the introduction and withdrawal is done via at fixed positions via multiport-valves. The connection of the columns with each other and to the inlets and outlets is very flexible, so that a counter-current operation like in an SMB or a co-current stream similar to annular chromatography can be realized. Figure 1.15 shows the carrousel principle

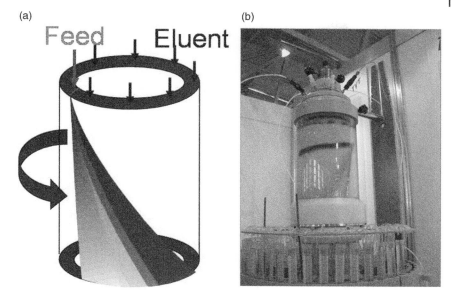

Figure 1.14 Annular chromatography (a) principle, (b) photo.

Figure 1.15 Carrousel absorbers (CSEP®).

with a resin volume from 1 liter to 300 m³. The plants are used especially in food industry in large dimensions (carrousel diameters of several meters).

However, there exist other niche techniques like, for example, circular chromatography or annular electrochromatography, which offer high numbers of separation stages [47, 48], but are very limited in throughputs and only applicable to high value pharmaceutical applications.

References

1. Levey, M. (1959) *Chemistry and Chemical Technology in Ancient Mesopotamia*, Elsevier, Amsterdam, The Netherlands.
2. Holmyard, E.J. (1957) *Alchemy*, A348 Pelican Books, Harmondsworth, Middx, UK.
3. Blaß, E., Liebl, T. and Häberl, M. (1997) *Chem. Ing. Tech.*, **69**, 431.
4. Ebeling, E. (1949) *Orientalia*, **17**, 129.
5. Ebeling, E. (1949) *Orientalia*, **18**, 404.
6. Ebeling, E. (1949) *Orientalia*, **19**, 265.
7. Joachim, H. (1973) *Papyros Ebers: Das älteste Buch über Heilkunde*, W. de Gruyter-Verlag, Berlin.
8. Cannell, R.J.P. (1990) How to approach the isolation of a natural product, in *Natural Products Isolation*, in Methods in Biotechnology, vol. 4 (ed. R.J.P. Cannell), Humana Press Inc., Totowa, NJ, USA, pp. 1–52.
9. http://en.wikipedia.org/wiki/natural_products (accessed 3 March 2009).
10. El-Mansi, E.M.T. (ed.) (2007) *Fermentation Microbiology and Biology*, 2nd edn, CRC Press, Boca Raton, USA.
11. Keller, K., Knöss, W., Reh, K. and Schnädelbach, D. (2003) Phytopharmaka – Begriffsbestimmungen und Hintergründe. *Bundesgesundheitsbl-Gesundheitsforsch-Gesundheitsschutz*, **46**, 1036. doi: 10.1007/s00103-003-0735x
12. Pzyrembel, H. (2003) Arzneipflanzen in Nahrungsergänzungsmitteln. *Bundesgesundheitsbl-Gesundheitsforsch-Gesundheitsschutz*, **46**, 1074. doi: 10.6007/s00103-003-0738.7
13. Bäcker, W. et al. (eds.) 2004, http://www.processnet.org/_positionspapiere.html (accessed 1 February 2011).
14. Bart, H.-J., Bischoff, F., Ulber, R. and Schmidt, M. (2006) Vom Naturstoff zum Wertstoff. Paper at "Grüne Woche Rheinland-Pfalz," 17.-20.1.2006, Ingelheim, Germany.
15. Mahlberg, B. (2009) DLR-Rheinpfalz, personal communication.
16. Bundesministerium für Gesundheit und Soziale Sicherung (ed.) (2009) *Deutsches Arzneibuch 2009*, (DAB 2009), 6. Ausgabe Grundwerk 2008 mit Nachtrag 6.1 und 6.2. 2009. Amtliche deutsche Ausgabe Deutscher Apotheker Verlag, Stuttgart, Germany.
17. Zöphel, B. and Kreuter, T. (2001) *Nachwachsende Rohstoffe (Hanf, Flachs, Salbei und Kamille)-Anbau und Betreuung für den Lebensraum Acker in Sachsen*, Sächs. Landesamt für Umwelt und Geologie, Dresden, Germany, pp 1–64.
18. Extraktionslösemittelverordnung, B.G.B.L., II Nr. 465/1998 (1998) Bundeskanzleramt, Vienna.
19. Lack, E. and Simandi, B. (2001) Super fluid extraction and fractionating from solid materials, in *High Pressure Process Technology: Fundamentals and Applications* (eds A. Bertucco and G. Vetter), Elsevier, Amsterdam, The Netherlands, pp. 537–575.
20. European Medicines Agency, (2006) ICH Topic Q3C (R3), Impurities: Residual Solvents, (March 1998 CPMP/IHC/382/95) http://www.ema.europa.eu/docs/en_GB/document_library/Scientific_guideline/2009/09/WC500002674.pdf (accessed 3 March 2009).
21. Jones, W.P. and Kingkorn, A.D. (2006) Extraction of plant secondary metabolites, in *Natural Products Isolation*, (eds S.D. Sarker, Z. Latif and A.I. Gray), Methods in Biotechnology, Vol. 20, 2nd edn, Humana Press, Totowa, NJ, USA, pp. 323–351.
22. Van Laar, J.J. (1990) *Z. Phys. Chem.*, **72**, 723.
23. Hildebrand, J.M. and Scott, R.L. (1950) *The Solubility of Nonelectrolytes*, 3rd edn, Reinhold, New York, USA.
24. Barton, A.F.M. (1991) *Handbook of Solubility Parameters and Other Cohesion Parameters*, 2nd edn, CRC Press, Boca Raton, USA.
25. Hansen, C. (2007) *Hansons Solubility Parameters, A User's Handbook*, 2nd edn, CRC Press, Boca Raton, USA.
26. Fedors, R.F. (1974) *Polym. Eng. Sci.*, **14**, 2, 147.
27. Kamlet, M.J., Abont, J.L.M. and Taft, R.W. (1981) *Prog. Phys. Org. Chem.*, **13**, 483.

28 Kamlet, M.J., Abont, J.L.M., Abraham, M.M. and Taft, R.W. (1983) *J. Org. Chem.*, **48**, 2877.

29 Dahlstrom, D.A., *et al.* (1997) Liquid-solids operation and equipment, in *Perry's Chemical Engineer's Handbook*, 7th edn (eds R.H. Peny and D.W. Green), Mc Graw Hill, New York, USA, pp. 1–59.

30 Bart, H.-J. and Schmidt, M. (2007) *Chem. Ing. Tech.*, **79**, 5, 663.

31 Eggers, R. and Jaeger, R.T. (2003) Extraction systems, in *Extraction Optimization in Food Engineering* (eds C. Tzia and G. Liadakis), Marcel Dekker, New York, pp. 95–136.

32 Savova, M., Bart, H.-J. and Seikova, I. (2005) *J. Univ. Chem. Techn. Metallurgy*, **40**, 251.

33 Cseke, L.J., *et al.* (ed.) (2006) *Natural Products from Plants*, 2nd edn, CRC Press, Boca Raton, USA.

34 Ganeva, V., *et al.* (2004) *Biotechnol. Lett.*, **26**, 933.

35 Coulson, J.M., Richardson, J.F., Backhurst, J.R. and Harker, J.H. (1995) *Chemical Engineering*, vol. 2, 4th edn, Pergamon Press, Oxford, UK.

36 Soxhlet, F. (1879) *Polytechnisches J.*, (Dingler's) **232**, 461.

37 Hunter, T.G. and Nash, A.W. (1934) *J. Chem. Soc.*, 53.

38 Marr, R. and Bart, H.-J. (1992) *Chem. Ing. Tech.*, **54**, 119.

39 Ladisch, M.L. (2001) *Bioseparation Engineering, Principles, Practice and Economics*, Wiley Interscience, New York, USA.

40 Schulte, M., Wekenbourg, K. and Strube, J. (2007) Continuous chromatography in the downstream processing of products of biotechnology and natural origin, in *Bioseparation and Bioprocessing*, vol. 61, 2nd edn (ed. G. Subramanian), Wiley-VCH Verlag GmbH, Weinheim, Germany, pp. 225–255.

41 Guichon, G. (2002) Basic principles of chromatography, in *Ullmann's Encyclopedia of Industrial Chemistry*, 6th edn, Wiley-VCH Verlag GmbH, Weinheim, Germany, pp. 1–31.

42 Sorti, G., Mazzotti, M., Morbidelli, M. and Carra, S. (1993) *AIChE J.*, **39**, 471.

43 DeCarli, J.P., II, Carta, G. and Byers, C.H. (1990) *AIChE J.*, **36**, 1220.

44 Bart, H.-J., Wolfgang, J., Prior, A. and Byers, C. (1998) *Chem. Ing. Tech.*, **70**, 142.

45 Reissner, K., Prior, A., Wolfgang, J. and Bart, H.-J. (1997) *J. Chromatogr. A*, **763**, 49.

46 Calgon Carbon Corporation, (2009) Ion Exchange Technologies. http://www.calgoncarbon.com/ion_exchange/index.html (accessed 29 November 2010).

47 Bart, H.-J. (2009) Vorrichtung für kont. Kapillarelektrochromatographie. German Patent Application, 102009013899.4.

48 Barker, P.E. (1966) *Brit. Chem. Engng.*, **11** (3), 203.

2
Solubility of Complex Natural and Pharmaceutical Substances

Feelly Ruether and Gabriele Sadowski

2.1
Introduction

The knowledge of solubility data is essential for high-purity separations (e.g., crystallization) and extraction processes. Due to low water solubility of natural and pharmaceutical compounds, other solvents (either as alternative solvents or co-solvents, respectively) are used in crystallization processes. The application of supercritical fluids (mostly carbon dioxide) for the extraction of natural products had also been investigated (e.g., [1, 2]). For designing such an extraction process the solubility and its dependence on temperature and pressure has to be known. However, it is practically impossible to measure the solubility of a compound at all conditions of temperature, pressure, and in all possible solvents and solvent mixtures. That is why a thermodynamic model which allows for the prediction of solubility data in different solvents and solvent mixtures at varying temperatures and pressures is of great interest.

2.2
Solubility Calculations

2.2.1
Solubility of a Pure Solute in Solvents and Solvent Mixtures

Several empirical and semi-empirical equations for solubility calculations can be found in the literature. The coefficients of the equations are considered either as functions of physical properties of the solute compounds and the solvents, such as polarity [e.g., within the QSPR (quantitative structure property relationship) method, as reported among others by Rytting *et al.* [3] and Huuskonen *et al.* [4]), or as functions of different group contributions [5], and molecular structures [6]. This kind of calculation generally requires a huge number of experimental solubility data to identify the respective model coefficients. Moreover, the application of

Industrial Scale Natural Products Extraction, First Edition. Edited by Hans-Jörg Bart, Stephan Pilz.
© 2011 Wiley-VCH Verlag GmbH & Co. KGaA. Published 2011 by Wiley-VCH Verlag GmbH & Co. KGaA.

these equations is mostly limited to those systems for which the experimental data were used for fitting of the coefficients.

From the thermodynamic point of view, the solubility can be calculated based on the phase-equilibrium condition for solid and liquid phases, that is, based on the equality of the chemical potentials of the solute compound (i) in both, solid and liquid phase:

$$\mu_i^L = \mu_i^S \tag{2.1}$$

Based on that, assuming a pure solid phase and neglecting the difference of solid and liquid heat capacities, the solubility of the solute compound i can be expressed by the following equation:

$$x_i^L = \frac{\varphi_{0i}^L(T,p)}{\varphi_i^L(T,p,x_i^L)} \cdot \exp\left[-\frac{\Delta h_{0i}^{SL}}{RT}\left(1 - \frac{T}{T_{0i}^{SL}}\right) - \frac{(v_{0i}^S - v_{0i}^L) \cdot (p_0 - p)}{RT}\right] \tag{2.2}$$

Here, x_i^L represents the solubility of the solute compound i (mole fraction in the liquid phase). φ_{0i}^L and φ_i^L are the fugacity coefficients of the solute compound i as a pure liquid and in the liquid mixture, respectively. The latter consists of the dissolved solute compound i and the solvent(s). The fugacity coefficients are thermodynamic quantities depending on temperature T and pressure p and are commonly calculated from equations of state. Δh_{0i}^{SL} and T_{0i}^{SL} are the melting enthalpy and the melting temperature of the pure solute compound i, respectively, whereas R is the gas constant. v_{0i}^S and v_{0i}^L are the molar volumes of the pure solute compound as solid (S) and liquid (L), respectively. p_0 is the standard pressure of 1 bar. Since the molar volume of the solid is not experimentally available and the difference between v_{0i}^S and v_{0i}^L is commonly very small, the second term in the exponent can be neglected in the solubility calculations, so that the pressure dependence of the solubility is only taken into account by the fugacity coefficients.

The ratio of the fugacity coefficient of component i in the liquid mixture and the fugacity coefficient in the pure liquid state is usually defined as the activity coefficient:

$$\gamma_i^L \equiv \frac{\varphi_i^L(T,p)}{\varphi_{0i}^L(T,p,x_i^L)} \tag{2.3}$$

Hence, Equation 2.2 finally reads as:

$$x_i^L = \frac{1}{\gamma_i^L(T,p,x_i^L)} \cdot \exp\left[-\frac{\Delta h_{0i}^{SL}}{RT}\left(1 - \frac{T}{T_{0i}^{SL}}\right)\right] \tag{2.4}$$

From Equation 2.4 it can be seen that the solubility of a solute—besides its pure-component properties (melting enthalpy and melting temperature)—only depends on its activity coefficient in solution. Therefore, the activity coefficient is the only quantity in the equation that is able to express the varying solubility of one solute in different solvents or solvent mixtures. The activity coefficient describes to what extent a solute molecule behaves differently in the liquid solution compared with the pure liquid solute. From Equation 2.3 it is obvious, that the activity coefficient is equal to one for a pure component or systems where solute and solvent have very similar/identical properties. For a mixture of unlike molecules, the activity coefficients differ from one. In case of complex solute/solvent systems, the molecules are definitely different from each other with respect to volume, polarity, etc. Thus, the activity coefficients in mixtures are almost never unity and hence can not be neglected in the solubility calculations.

The activity coefficients can be directly calculated using activity-coefficient models as for example, NRTL-SAC [7] and COSMO-RS [8]. However, these models do neglect the pressure-dependence of that quantity and thus cannot be applied to predict or to correlate the solubility in supercritical fluids at elevated pressures. Therefore, in this work we apply a more general approach and calculate the fugacity coefficients from an equation of state (PC-SAFT [9]) which are used to obtain the activity coefficients according to Equation 2.4.

2.2.2
pH-Dependence of Solubility

For amino acids in aqueous electrolyte solution there exist three characteristic equilibrium reactions: the formation of a zwitterion, the protonation of the carboxyl group, and deprotonation of the amino group, which can be characterized by the corresponding equilibrium constants. A simple approach to account for pH effects on the solubility of amino acids ($x^L_{AA,total}$ in Equation 2.5) can be derived from the knowledge of the solubility of an uncharged amino acid (e.g., at its isoelectrical point, $x^L_{AA^\pm}$ in Equation 2.5) and the two equilibrium constants K_{A1} and K_{A2} [10]:

$$x^L_{AA,total} = x^L_{AA^\pm} \cdot \left(1 + \frac{10^{-pH}}{K_{A1}} + \frac{K_{A2}}{10^{-pH}}\right) \quad (2.5)$$

The values of K_{A1} and K_{A2} at different temperatures are obtained from the van't Hoff equation based on the knowledge of the equilibrium constants at $T_0 = 25\,°C$ as well as the corresponding enthalpies of protonation:

$$K_A(T) = K_A(T_0) \cdot \exp\left(\frac{\Delta h_{protonation}}{R} \cdot \left(\frac{1}{T} - \frac{1}{T_0}\right)\right) \quad (2.6)$$

2.2.3
Solubility of Racemic Compounds

In order to separate pure enantiomers from a racemic mixture it is essential to know the phase equilibrium of those systems. Whereas the solubility of the pure enantiomers can be modeled using Equation 2.4, the solubility of the racemic compound can be modeled applying a quasi-chemical approach. Given that the racemic compound exists in the solid phase only, its solubility can be calculated by treating it as the result of a chemical reaction between the two liquid enantiomers (here e.g., the enantiomers (+)MA and (−)MA of racemic mandelic acid (rac. MA)):

$$(+)MA^L + (-)MA^L \Leftrightarrow rac.\ MA^S \tag{2.7}$$

Assuming that the solid racemic compound precipitates as a pure solid, that is, the activity of this compound is equal to unity, the equilibrium constant for the Reaction (2.7) can be expressed as a function of the activity coefficients of the two liquid enantiomers according to

$$K_a = \prod_i a_i^{v_i} = \left(x^L_{(+)MA} \cdot \gamma^L_{(+)MA}\right)^{-1} \cdot \left(x^L_{(-)MA} \cdot \gamma^L_{(-)MA}\right)^{-1}$$

$$= \left(x^L_{(+)MA} \cdot \gamma^L_{(+)MA}\right)^{-1} \cdot \left(\left(1 - x^L_{(+)MA}\right) \cdot \gamma^L_{(-)MA}\right)^{-1} \tag{2.8}$$

Furthermore, the equilibrium constant K_a at different temperatures equals to

$$\ln K_a = \ln K_{a,rac.MA} + \frac{\Delta h^{SL}_{rac.MA}}{R} \cdot \left(\frac{1}{T^{SL}_{rac.MA}} + \frac{1}{T}\right) \tag{2.9}$$

Therein, $\Delta h^{SL}_{rac.MA}$ and $T^{SL}_{rac.MA}$ are the melting enthalpy and the melting temperature of the pure racemic compound (here: mandelic acid), respectively. $K_{a,rac.MA}$ is the value of the equilibrium constant calculated at $T^{SL}_{rac.MA}$. The solubility calculation of racemic compound is performed by combining Equations 2.8 and 2.9. The activity coefficients in Equation 2.8 are determined in the same way as for the solubility calculation of the pure enantiomers (according to Equation 2.4).

2.3
Thermodynamic Modeling

2.3.1
PC-SAFT Equation of State

The activity coefficient of a solute compound i in the mixture is estimated in this work via the PC-SAFT equation of state. Using this model, the residual Helmholtz energy A^{res} of a system containing the solute compound and solvent(s) is considered as the sum of different contributions resulting from repulsion (hc: hard chain), van der Waals attraction (disp: dispersion) and hydrogen bonding (assoc: association) according to:

$$A^{res} = A^{hc} + A^{disp} + A^{assoc} \tag{2.10}$$

The different contributions to the Helmholtz energy according to PC-SAFT as well as the model parameters are briefly presented below. The detailed equations for each contribution and the equations which relate the activity coefficients to the residual Helmholtz energy can be found in the appendix.

2.3.1.1 Hard-Chain Contribution A^{hc}
The hard-chain reference fluid consists of spherical segments which do not exhibit any attractive interactions. To define a hard-chain fluid, two parameters are

required, namely the number of segments m and the segment diameter σ. The Helmholtz energy of this reference system is described by an expression developed by Chapman et al. [11] which is based on Wertheim's first order thermodynamic perturbation theory [12–14].

To calculate mixture properties (e.g., solute compound i in solvent j), the segment diameters of the two pure components are combined via conventional Berthelot–Lorentz rules according to

$$\sigma_{ij} = \frac{1}{2}(\sigma_i + \sigma_j) \tag{2.11}$$

and the mean segment number in the mixture $\bar{m}$ is calculated as

$$\bar{m} = \sum_i x_i m_i \tag{2.12}$$

2.3.1.2 Dispersion Contribution A^{disp}

The contribution of dispersive attractions (correspond to van der Waals interactions) to the Helmholtz energy of a system according to PC-SAFT was derived from the perturbation theory of Barker and Henderson [15, 16] applied to the hard-chain reference system. In addition to the above-mentioned parameters, one additional parameter is required for describing the segment-segment interaction of different molecules, which is the dispersion energy parameter ε/k.

To calculate the mixture property from pure-component parameters, again conventional Berthelot–Lorentz combining rules are applied:

$$\left(\frac{\varepsilon}{k}\right)_{ij} = \sqrt{\left(\frac{\varepsilon}{k}\right)_i \cdot \left(\frac{\varepsilon}{k}\right)_j} \cdot (1 - k_{ij}) \tag{2.13}$$

Equation 2.13 also contains one adjustable binary interaction parameter k_{ij} which is used to correct the dispersion energy in the mixture. The value of k_{ij} between solute and solvent is determined from fitting the k_{ij} to binary solubility data. The values of k_{ij} between different solvents were fitted to the binary vapor-liquid equilibria and have already been partly reported in the literature (e.g., [17]). For the description of ternary or multi-component systems it is assumed that two-molecule interactions dominate and thus no other than binary parameters are required.

2.3.1.3 Association Contribution A^{assoc}

The contribution due to short-range association interactions (hydrogen bondings) A^{assoc} is considered by an association model which was proposed by Chapman et al. [18] based on Wertheim's first-order thermodynamic perturbation theory (TPT1). Within this theory a molecule is assumed to have one or more association sites (electron acceptors or electron donors) which can form hydrogen bonds. The association between two different association sites of the same molecule i is characterized by two additional parameters: the association strength $\varepsilon^{A_i B_i}/k$ and the effective volume of the association interaction $\kappa^{A_i B_i}$. Hence, an associating compound is characterized by a total of five pure-component parameters (segment number m, segment diameter σ, dispersion energy parameter ε/k, association strength $\varepsilon^{A_iB_i}/k$, and the effective volume of the association interaction $\kappa^{A_iB_i}$.

The strength of cross-association interactions between the association sites of two different compounds (A_i and B_j) can be determined using simple combining rules of the pure-component parameters as suggested by Wolbach and Sandler [19] without introducing any further binary parameters:

$$\frac{\varepsilon^{A_iB_j}}{k} = \frac{1}{2}\left(\frac{\varepsilon^{A_iB_i}}{k} + \frac{\varepsilon^{A_jB_j}}{k}\right) \quad (2.14)$$

$$\kappa^{A_iB_j} = \sqrt{\kappa^{A_iB_i} \cdot \kappa^{A_jB_j}} \cdot \left(\frac{\sqrt{\sigma_i \cdot \sigma_j}}{\frac{1}{2}(\sigma_i + \sigma_j)}\right)^3 \quad (2.15)$$

Knowing the molecular structure of a molecule, one can choose the terms that have to be included in the calculation of the Helmholtz energy. Nonpolar or weakly-polar components can be satisfactorily described using only the hard-chain and dispersion contributions. For associating compounds, the association contribution also needs to be considered. For example, for estriol, see Figure 2.1, two different types of association sites are assumed (electron acceptor and electron donor), each of them existing three times per molecule. The two association types were assumed to be of equal strength and range. Thus, only one value for the association strength ($\varepsilon^{A_iB_j}/k$) and one value for the association volume ($\kappa^{A_iB_j}$) parameter need to be determined.

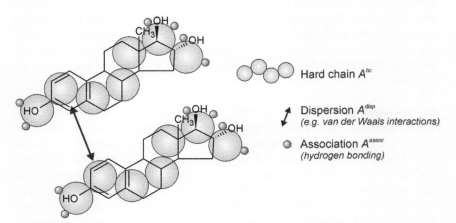

Figure 2.1 Molecular model of estriol within the framework of PC-SAFT.

2.3.2
Estimation of PC-SAFT Parameters

For solvents (liquids and gases), equation-of-state parameters are best fitted to liquid densities and vapor pressures of the pure components [9]. However, under normal conditions most of the natural and pharmaceutical products exist only as solids and neither vapor-pressure data nor liquid densities are available. Hence, another approach to determine the pure-component parameters for the solute compound is applied: the parameters for the solutes are determined by fitting them to a small number of solubility data in one pure solvent.

Consequently, the required amount of experimental solubility data depends on the number of parameters (three or five pure-component parameters) which have to be adjusted for the solute. As binary solubility data are used to determine the solute parameters, the k_{ij} between solute and solvent has to be estimated simultaneously. The number of adjustable parameters can often be reduced by setting the parameter $\kappa^{A_iB_i}$ to a value between 0.01 and 0.03, which are common values for association volumes for most organic substances.

Moreover, the melting enthalpy and melting temperature need to be known as indicated in Equation 2.4. It is recommended that experimental melting enthalpy and melting temperature values are used, which can be determined independently by differential scanning calorimetry (DSC). If melting temperature and melting enthalpy are unknown, and for solutes which decompose before or during melting (e.g., amino acids), both the melting temperature and the melting enthalpy can also be treated as adjustable parameters [20].

When dealing with polymorphs, Equation 2.4 can also be applied by using different melting enthalpies and melting temperatures for the different polymorphs.

If the solubility data in one solvent are not accessible at various temperatures, one can also use solubility data of the solute in several solvents. In order to reduce the number of parameters to be fitted, it is suggested that the solvents be divided into certain classes, according to their molecular structures and physical properties; for example, nonpolar, weakly polar, strongly polar, and associating components since the k_{ij} values between one solute and all solvents belonging to the same class can be assumed to be identical [21].

A list of pure-component parameters of some natural and pharmaceutical compounds obtained in this way is presented in Table 2.1. The table also contains the information which solvent was used to determine the pure-component parameters, as well as the melting temperatures and melting enthalpies used in the calculations. The details of the parameter-determination procedures can be found in the next section. For the sake of completeness, the pure-component parameters for the solvents considered here are given in the Table 2.1 as well. Table 2.2 lists the determined k_{ij} parameters for each of the solute–solvent and solvent–solvent systems considered.

Table 2.1 Pure component PC-SAFT parameters and melting properties for considered natural and pharmaceutical products as well as the used solvents.

Compounds	Segment number m [-]	Segment diameter σ [Å]	Energy parameter (dispersion) ε/k [K]	Energy parameter (association) $\varepsilon^{A_iB_i}/k$ [K]	Association volume $\kappa^{A_iB_i}$ [-]	Number of electron acceptors [-]	Number of electron donors [-]	Solvent(s) for parameter estimation	Melting temperature T_{0i}^{SL} [°C]	Melting enthalpy Δh_{0i}^{SL} [J/mol]
Solutes										
Estriol	7.532	4.438	471.825	2910.238	0.03	3	3	acetone	282	41 870
Estrone	7.553	3.288	298.446	1170.515	0.03	2	2	acetone	260.2	34 480
Sitosterol	12.548	2.883	265.984	2817.570	0.03	1	1	acetone	140	42 080
Beta-carotene	10.719	3.483	198.211	0	0	0	0	hexane, cyclohexane	183	20 802
Paracetamol	7.524	3.508	398.284	1994.228	0.01	2	2	water	170.45	27 000
DL-methionine	8.712	2.737	237.181	1964.000	0.033	2	2	water	826.85	28 600
(+)-/(-)-mandelic acid	7.608	3.334	407.209	2093.614	0.01	2	2	water	131.5	24 500
Solvents/low-molecular compounds										
Acetone	2.891	3.228	247.418	0	0.03	1	1	–	–	–
Ethanol	2.383	3.177	198.237	2653.384	0.032	1	1	–	–	–
Chloroform	2.589	3.442	266.381	0	0.01	1	1	–	–	–
Toluene	2.815	3.717	285.690	–	–	–	–	–	–	–
Benzene	2.465	3.648	287.355	–	–	–	–	–	–	–
Dichloromethane	2.589	3.278	266.434	–	–	–	–	–	–	–
Hexane	3.058	3.798	236.769	–	–	–	–	–	–	–
Cyclohexane	2.530	3.850	278.109	–	–	–	–	–	–	–
Carbon dioxide	2.073	2.785	169.207	–	–	–	–	–	–	–
Water	1.066	3.001	366.512	2500.671	0.035	1	1	–	–	–
2-propanol	3.093	3.209	208.420	2253.9	0.025	1	1	–	–	–
1-butanol	2.752	3.614	259.591	2544.560	0.01	1	1	–	–	–

Table 2.2 Binary parameters for solute/solvent and solvent/solvent interactions.

System	Binary parameter k_{ij}
Solute/solvent	
Estriol/acetone	$-0.00002470*T$ [K] $- 0.00785127$
Estriol/ethanol	$0.00065250*T$ [K] $- 0.19973038$
Estriol/chloroform	$0.00009168*T$ [K] $- 0.04061030$
Estriol/toluene	$0.00032561*T$ [K] $- 0.18606360$
Estriol/benzene	$0.00056874*T$ [K] $- 0.25300143$
Estriol/dichloromethane	$0.00034683*T$ [K] $- 0.15054061$
Estrone/acetone	$0.00028473*T$ [K] $- 0.11356383$
Sitosterol/acetone	$0.00031569*T$ [K] $- 0.12173125$
Beta-carotene/hexane	$-0.00106888*T$ [K] $+ 0.37528091$
Beta-carotene/cyclohexane	$0.00098000*T$ [K] $- 0.21738700$
Beta-carotene/carbon dioxide	$0.00057113*T$ [K] $- 0.00818560$
Paracetamol/water	$0.00028276*T$ [K] $- 0.10072067$
Paracetamol/ethanol	$0.00021306*T$ [K] $- 0.13203682$
Paracetamol/acetone	$0.00026365*T$ [K] $- 0.13297859$
Paracetamol/toluene	$0.00065331*T$ [K] $- 0.24063338$
DL-methionine/water	$0.00011667*T$ [K] $- 0.13220083$
(+)-mandelic acid/water	$-0.00037813*T$ [K] $+ 0.09071922$
(−)-mandelic acid/water	$-0.00037813*T$ [K] $+ 0.09071922$
(+)-mandelic acid/(−)-mandelic acid	-0.065
Solvent/solvent	
Water–acetone	-0.055
Water–ethanol	0.02

2.4 Examples

In this section we present typical examples of solubility calculations. We distinguish between correlation and prediction. "Correlation" means that a few experimental data were used to identify the model parameters and to give a quantitative representation of the solubility data, whereas the term "prediction" is applied when the solubility curve of a solute compound in a solvent or solvent mixture is calculated without using any experimental data of the considered system.

2.4.1
Solubility of Estriol, Estrone, and Sitosterol in Different Solvents

Estriol, estrone and sitosterol were chosen as examples of complex natural molecules, for which literature data on solubilities in different pure solvents are available and summarized in [22, 23]. The melting temperatures were taken from the CAPEC Database [24] and the melting enthalpies were estimated with Marrero and Gani's method [25] as suggested in [22, 23].

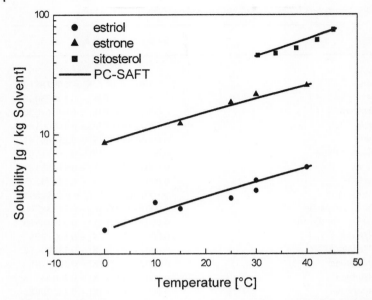

Figure 2.2 Solubility of estriol, estrone, and sitosterol in acetone. Comparison of experimental data from literature (symbols) with correlation results of PC-SAFT (lines).

Estriol was considered as an associating compound with three electron-acceptor and three electron-donor sites, respectively. Estrone was also modeled assuming two electron acceptors and two electron donors, respectively. For sitosterol, only two association sites (one electron acceptor and one electron donor) were assigned. To reduce the number of fitting parameters, the value of the association volume was set to 0.03 for all three components.

The experimental solubility data of estriol, estrone as well as sitosterol in acetone from [26–28] were used to fit the PC-SAFT pure-component parameters of the solutes and the binary parameters k_{ij} between each of the solutes and acetone. The results of the correlation are depicted in Figure 2.2.

Using a linear temperature dependency of the binary parameter k_{ij}, the correlation results can be brought into satisfactory agreement with the experimental data. The use of a linear temperature-dependent binary parameter k_{ij} can be justified for two reasons [21]:

a) The thermodynamic relationship to calculate the solubility according to Equation 2.4 neglects the influence of the heat capacities which are temperature dependent but in most of the cases not available experimentally. Therefore, k_{ij} would be expected to be temperature dependent to correct for this assumption.

b) The determination of the linear temperature function of k_{ij} requires only two data points. The linear function can afterwards be adopted for a safe

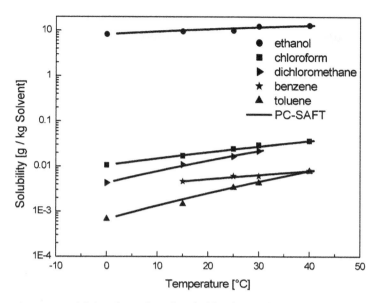

Figure 2.3 Solubility of estriol in ethanol, chloroform, toluene, benzene, and dichloromethane. Comparison of experimental data from literature (symbols) with correlation results of PC-SAFT (lines).

extrapolation to other temperatures which will be demonstrated later for example, in Figures 2.3 and 2.6b.

After determining the pure-component parameters for the solutes, one can use these parameters to estimate the solubility of the solutes in different solvents. In those cases, the only parameter that needs to be determined is the binary parameter k_{ij} between the solute and the respective solvent. If k_{ij} were to be set to zero, the calculation would be a pure prediction. However, for quantitative agreement, in most cases the binary parameter k_{ij} between two unlike molecules has to be fitted to at least one experimental data point. This is demonstrated here for the solubility calculation of estriol in ethanol, chloroform, toluene, benzene, and dichloromethane (Figure 2.3). Only two solubility data in each solvent are required to fit the corresponding binary parameter between estriol and the solvent. Using this approach, the PC-SAFT model is able to correlate the temperature dependence of the solubility in different solvents in good agreement with the experimental data.

2.4.2
Solubility of Beta-carotene in Supercritical Carbon Dioxide

Beta-carotene is chosen here as an example of natural products, for which the solubility in supercritical carbon dioxide at different temperatures and elevated

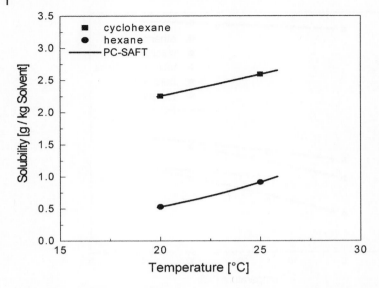

Figure 2.4 Experimental solubility data of beta-carotene in hexane and in cyclohexane from literature (symbols) and correlation results of PC-SAFT (lines).

pressures had been investigated in the literature (e.g., [29–31]). The molecular model of beta-carotene ($C_{40}H_{56}$) does not include association; hence only three pure-component parameters are required. The melting temperature was taken from [24]. No experimental or estimated value of the melting enthalpy was available in the literature. Hence, the melting enthalpy was treated as adjustable parameter like the pure-component parameters for beta-carotene which were fitted simultaneously to the solubility data [32, 33] in cyclohexane and hexane (a total of four data points) at atmospheric pressure. In the first step of the parameter fitting, the binary parameters between beta-carotene/cyclohexane and beta-carotene/hexane were assumed to be equal and the segment number was set to a value of around 10, to reduce the number of adjustable parameters. In the second step, the respective binary parameters and the segment number were readjusted to give quantitative correlation results for the solubility of beta-carotene in cyclohexane and in hexane (Figure 2.4).

To calculate the solubility of beta-carotene in supercritical carbon dioxide, the binary parameter between beta-carotene and CO_2 is the only unknown parameter left. For the determination of a linear temperature-dependent k_{ij}, only two solubility points are required. Here we used the data points at 150 bar and 34 °C as well as at 60 °C from [30, 34]. The experimental values (Figure 2.5a) show a plateau at about 0.00075 g/kg carbon dioxide. This nonlinear temperature dependence of the solubility is also captured by the model calculations giving a solubility minimum in the same range. Considering the extremely small absolute values of the solubility, it can be concluded the PC-SAFT is able to reproduce the temperature dependence of the solubility even using a linear temperature function for k_{ij}.

(a)

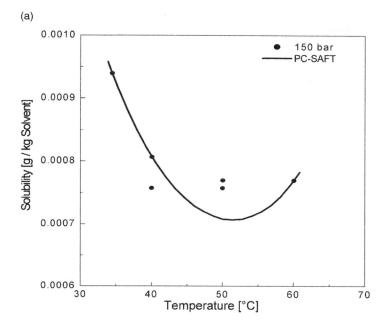

(b)

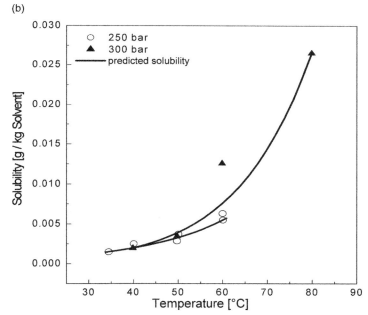

Figure 2.5 (a) Temperature dependence of the solubility curve of beta-carotene in carbon dioxide at 150 bar. Experimental data from literature (symbols) compared with correlation results of PC-SAFT (lines). (b) Predicted solubility of beta-carotene in carbon dioxide at 250 and 300 bar, respectively, by PC-SAFT (lines) in comparison with experimental data from literature (symbols).

Since all parameters have now been determined, the subsequent calculations at different pressures (250 bar and 300 bar) are therefore pure predictions, that is, no solubility data is required to produce the results which are shown in Figure 2.5b compared with experimental data [30, 31, 34, 35]. The PC-SAFT model is able to predict the solubility at those pressures without any further fitting to experimental data, which reduces the experimental effort tremendously.

2.4.3
Solubility of Paracetamol in Pure Solvents and Solvent Mixtures

Experimental values of the melting enthalpy (27 kJ mol^{-1}) and melting temperature of 170.45 °C from [36] were used for the solubility calculation of paracetamol in pure solvents. Paracetamol was modeled as an associating compound with two electron acceptors and two electron donors, respectively. The association strength between the electron acceptor and electron donor was assumed to be equal. The pure-component parameters of paracetamol – except for the association volume which was set to 0.01 – and the binary parameter k_{ij} for paracetamol/water were fitted to the experimental solubility data in water [36]. Figure 2.6a shows the result of the correlation (line) compared with experimental data (symbols). The PC-SAFT model is able to correlate the solubility of paracetamol in water; the results show a very good agreement (average absolute deviation for the temperature range =1.63%) with experimental data over the temperature range from 0 °C to 30 °C.

As mentioned above, to reduce the number of fitted parameters we suggest dividing the solvents into different classes and using the same k_{ij} between one solute and all solvents belonging to the same class. We have chosen three representatives: ethanol to represent the associating solvents, acetone for weakly-polar solvents, and toluene for aromatic solvents. The binary parameters k_{ij} between paracetamol and each of the solvents were fitted to only two experimental solubility points (0 °C and 30 °C, respectively). The results of the correlations are compared with experimental data at other temperatures as shown in Figure 2.6b. The PC-SAFT model can again remarkably well reproduce the temperature dependence of the paracetamol solubility in these different solvents for the considered temperature range.

To demonstrate the predictive capability of the PC-SAFT model, the solubilities of paracetamol in other pure alcohols (2-propanol and 1-butanol) are subsequently predicted using the paracetamol PC-SAFT parameters and the k_{ij}-value for paracetamol/ethanol. The latter means that all alcohols are considered here to belong to the same solvent class, and the binary parameter k_{ij} between paracetamol and different alcohols was set to the k_{ij}-value for paracetamol/ethanol. Hence, the solubility of paracetamol in 2-propanol and 1-butanol calculated this way is purely predicted, and the results are presented in Figure 2.7. Although not in quantitative agreement, these results are accurate enough for solvent-screening purposes in the process development. They clearly indicate that 2-propanol is the better solvent for paracetamol compared with 1-butanol. Thus, by applying the PC-SAFT model, the number of necessary screening experiments can be significantly reduced.

(a)

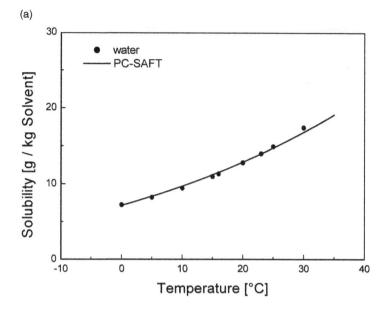

(b)

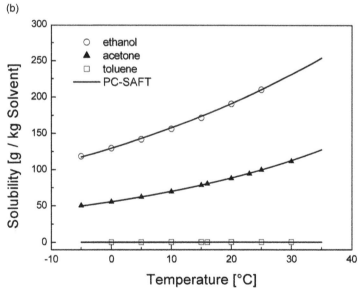

Figure 2.6 (a) Solubility data of paracetamol in water from literature (symbols) were used to fit the PC-SAFT parameters for paracetamol. The line is the result of the fitting. (b) PC-SAFT model is able to give the temperature dependence of the solubility curves of paracetamol in several solvents: ethanol, acetone, and toluene (lines) by using only two experimental solubility data points for each solvent (symbols).

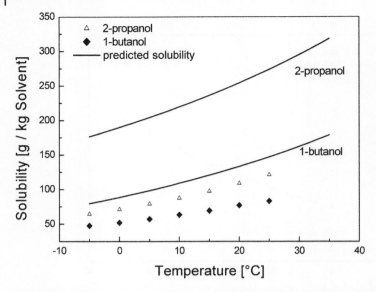

Figure 2.7 Predicted solubility curves of paracetamol in 2-propanol and in 1-butanol (lines) compared with experimental data from literature (symbols). PC-SAFT model is able to predict the influence of the different solvents qualitatively.

To increase the solubility of the pharmaceuticals, solvent mixtures or co-solvents are often used. However, it is practically impossible to scan all possible solvent mixtures and mixture compositions experimentally. Therefore, a thermodynamic model is needed to give qualitative information on the solubility in solvent mixtures based on the experimental solubility data in the pure solvents only.

The effect of the co-solvent on the solubility can show an unexpected behavior, as for example, shown for the solubility of paracetamol in water–acetone mixtures of various compositions at 25 °C (Figure 2.8a, symbols). The x-axis characterizes the amount of water in the water–acetone mixtures on a solute-free basis. The y-axis represents the solubility of paracetamol in the water–acetone mixtures of varying composition ranging from pure acetone to pure water. Although water is a worse solvent than acetone, the solubility of paracetamol in acetone can be almost quadrupled just by adding a small amount of water to the solution. One could have expected that water would act like an anti-solvent for paracetamol if added to acetone (decreasing solubility), but the opposite behavior is observed experimentally.

Using only the pure-component as well as binary parameters determined from the solubility in the pure solvents in the previous section, the solubility of pharmaceutical compounds in different solvent mixtures is subsequently predicted without fitting any additional parameters. This means that no extra experimental

(a)

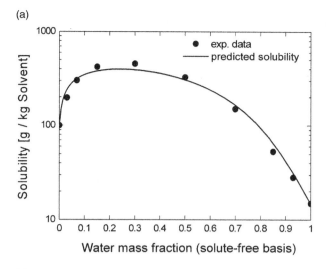

(b)

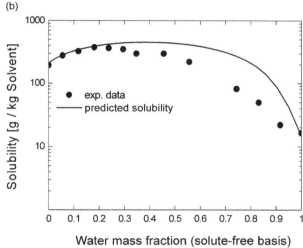

Figure 2.8 (a) Predicted solubility of paracetamol in water–acetone mixtures in the whole concentration range at 25 °C (line). No experimental solubility data of paracetamol in water–acetone mixtures (symbols) was required to predict the solubility curve. (b) Comparison of experimental solubility data of paracetamol in water–ethanol mixtures at 25 °C (symbols) with prediction results with PC-SAFT (line).

data is required to predict the solubility in the solvent mixtures in this work. The k_{ij}-values between the different solvents were for most cases already independently determined by fitting to binary vapor–liquid or liquid–liquid equilibria of the solvent mixtures, respectively, and can be found in the literature (e.g., in [17]).

Figure 2.8a also shows the predicted solubility of paracetamol in acetone–water mixtures at 25 °C (solid line). As it can be seen, the predicted curve gives a good description of the nonlinear dependency of the solubility on the solvent composition based only on the knowledge of the solubility of paracetamol in pure water and pure acetone, respectively.

Similar results were obtained for paracetamol in other mixed-solvent systems, for example, in water–ethanol mixtures which are shown in Figure 2.8b. The predicted solubility is in fair agreement with the experimental data without fitting any additional parameters.

2.4.4
Solubility of DL-Methionine as Function of pH

Lacking reliable data on melting temperatures and enthalpies for the amino acids, a hypothetical melting temperature of DL-methionine was fitted to the slope of the solubility curves in water, shown in Figure 2.9. The hypothetical melting temperature was estimated to 826.85 °C for DL-methionine [20] (this temperature should not be understood as the real physical property of the substance, since it is used here as an adjustable parameter). The melting enthalpy was estimated using the group-contribution method proposed by Marrero and Gani [25].

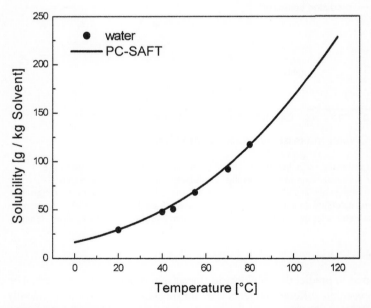

Figure 2.9 Solubility data of DL-methionine in water from literature (symbols) and correlation results of PC-SAFT (line).

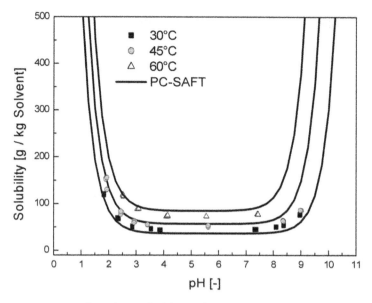

Figure 2.10 pH-dependence of solubility of DL-methionine at different temperatures (symbols: experimental data from literature at 30 °C, 45 °C, and 60 °C. Lines are calculation results of PC-SAFT using only the solubility at isoelectric point and pKa-values of DL-methionine.

To characterize the association of DL-methionine, two different types of association sites (each of them exists twice) were assumed. Both types were assumed to be of equal strength. The pure-component parameters for DL-methionine and the binary parameter k_{ij} between DL-methionine and water were fitted to the solubility data in water [20]. For DL-methionine, the PC-SAFT model is able to correlate the solubility in water and the result shows good agreement compared with experimental data (Figure 2.9).

Using the same model parameters as estimated above, also the pH-influence on solubility can be described using the information of the pKa-values of DL-methionine. Figure 2.10 shows the solubility of DL-methionine in different HCl/NaOH-solutions at 30 °C, 45 °C, and 60 °C. The symbols represent data from gravimetric measurements [20] and lines show the predicted solubility curves. Prediction results again show good agreement in the isoelectrical band and in this case in the acidic environment as well.

2.4.5
Solubility of Mandelic Acid Enantiomers and Racemic Mandelic Acid in Water

The schematic phase diagram of a system containing water, two enantiomers: (+)-mandelic acid, (−)-mandelic acid, and the racemic mandelic acid is shown in Figure 2.11. It is obvious that knowledge of the solubility and hence of the detailed

46 | *2 Solubility of Complex Natural and Pharmaceutical Substances*

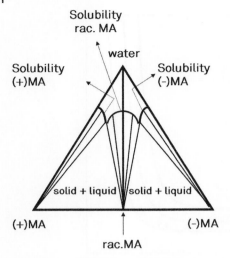

Figure 2.11 Schematic phase diagram of system containing water, (+)-mandelic acid, (−)-mandelic acid, and racemic mandelic acid. The solubility curves of each species are identified as well as the solid-liquid regions.

phase diagram is crucial for designing the crystallization process for separation and purification of the particular species.

The challenge for the modeling here is to describe the whole phase diagram using only one set of parameters and the small amount of experimental data required to identify the parameters. For the system mandelic acid in water, the melting enthalpies of the pure mandelic acid (+)- and (−)-enantiomers as well as of the racemic mandelic acid were reported in [37]. Also, the melting temperatures of the pure mandelic acid (+)- and (−)-enantiomers as well as of the racemic mandelic acid in water were also tabulated. Therein, the solubility data of all the mandelic acid species in water at different temperatures were presented.

The modeling procedure was performed as followed: since all the experimental values of the melting temperatures and melting enthalpies of the mandelic acid enantiomers as well as of the racemic mandelic acid are available, only the pure-component parameters of the (+)-mandelic acid, (−)-mandelic acid, as well as the binary parameters between (+)-mandelic acid/water, (−)-mandelic acid/water, and (+)-mandelic acid/(−)-mandelic acid need to be identified. Due to the fact that (+)-mandelic acid and (−)-mandelic acid are indistinguishable in the pure liquid phase, the pure-component parameters for (+)-mandelic acid and (−)-mandelic acid should be identical. Hence, the binary parameter for (+)-mandelic acid/water should also be the same as the binary parameter for (−)-mandelic acid/water. It is also obvious, that the presence of (−)-mandelic acid in the aqueous solution of

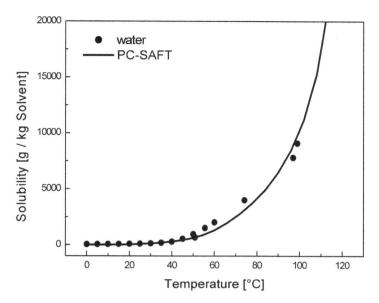

Figure 2.12 Solubility of (+)-mandelic acid in water as a function of temperature. Symbols are experimental data from literature and line is correlation results of PC-SAFT.

(+)-mandelic acid has a significant influence on the solubility of (+)-mandelic acid and vice versa. Therefore, the binary parameter for (+)-mandelic acid/(−)-mandelic acid cannot be neglected.

The pure-component parameters for the mandelic acid enantiomer as well as the binary parameter between mandelic acid enantiomer and water were fitted to the solubility data of (+)-mandelic acid in water (Figure 2.12).

To calculate the solubility of (+)-mandelic acid in water with a small amount of (−)-mandelic acid, the only parameter which is still unknown is the binary parameter for (+)-mandelic acid/(−)-mandelic acid. This parameter (a constant) was fitted to one solubility point of the ternary system at 35 °C. The comparison of the calculated solubility curve with the experimental data is depicted in Figure 2.13.

The solubility of the racemic compound in the ternary liquid solution (water, (+)-mandelic acid, and (−)-mandelic acid) was also predicted providing the knowledge of the solubility of the pure racemic compound in water. Figure 2.14 shows the comparison of the predicted results with the experimental data of the racemic-compound solubility at 35 °C, as well as the completed phase diagram for the system mandelic acid in water at 35 °C. It can be seen that PC-SAFT is able to reproduce the solubility curves of all mandelic acid species in a very good agreement to experimental data.

48 | *2 Solubility of Complex Natural and Pharmaceutical Substances*

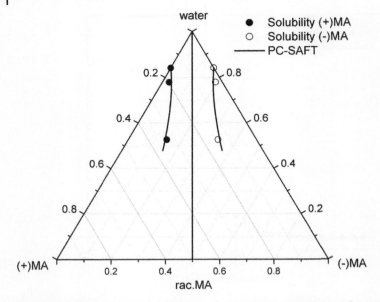

Figure 2.13 Symbols: solubility data of (+)-mandelic acid in the presence of water and (−)-mandelic acid as well as solubility data of (−)-mandelic acid in the presence of water and (+)-mandelic acid from literature. Lines are correlation results of PC-SAFT using only one solubility point in the ternary system. Solubilities are shown in mass fraction.

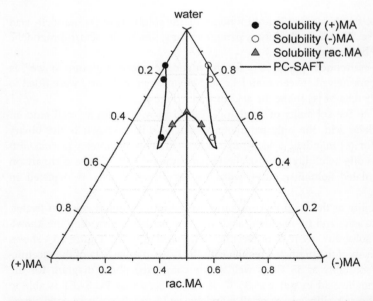

Figure 2.14 The complete calculated phase diagram (in mass fraction) with PC-SAFT (lines) for the system water, (+)-mandelic acid, (−)-mandelic acid, and racemic mandelic acid. The symbols are experimental solubility data of each mandelic-acid species from literature.

2.5
Summary

The solubility of several natural and pharmaceutical compounds in different pure solvents as well as in various solvent mixtures was modeled and predicted using the PC-SAFT model. When available, independently determined experimental melting temperatures and melting enthalpies were used in the calculations. To identify the PC-SAFT parameters for the solute, only a few solubility data in one pure solvent are required. For an adequate quantitative estimation of the solute's solubility in other solvents, it is recommended that the solvents be divided into certain classes according to their molecular structures and physical properties and to measure two solubility points for one solvent in each solvent class.

The PC-SAFT model allows for an accurate description of the temperature dependence of the solubility by applying a linear temperature-dependent interaction parameter for each binary system. The solubility in different solvents in the same solvent class can be estimated without any extra effort of experiments.

Based on the modeling results of the solubility in pure solvents, the solubility in mixed solvents can subsequently be reliably predicted without any need of additional fitting of parameters, hence again without extra experiments. The prediction results of solubility in solvent mixtures are in fair to very good agreement with the experimental data for all considered solvent mixtures and solvent-mixture compositions.

The model can also be reliably applied for calculations of solubility at elevated pressures. The pressure dependence is included in the thermodynamic model. Thus, no extra parameter is needed which would require more experimental solubility data. It can be demonstrated that the extrapolation to other pressures can be safely performed to reduce the amount of experimental data.

The pH-dependence of the solubility of amino acids can simply be taken into account using the knowledge of the pK_a-values of the respective amino acid. By applying the PC-SAFT model, the solubility at different temperatures and pH-values can be correlated and predicted as well.

Furthermore, an approach to calculate the solubilities of enantiomers and racemic compound in solvent or solvent mixtures using the same set of parameters for the whole phase diagram was presented. It can be shown exemplarily for the system of mandelic acid and water that the PC-SAFT model can successfully describe the solubility of the enantiomer species and the racemic compound using only a small amount of experimental data to identify the model parameters.

Symbols

Latin Symbols

a_i activity of component i [-]
A Helmholtz energy [J]

h	enthalpy [J mol^{-1}]
k	Boltzmann constant [J K^{-1}]
K	equilibrium constant [-]
k_{ij}	binary interaction parameter [-]
m	number of segments per molecule [-]
$\bar{m}$	mean segment number [-]
N	total number of molecules [-]
N_{AV}	Avogadro number [mol^{-1}]
p	pressure [Pa]
R	gas constant [J mol^{-1} K^{-1}]
T	temperature [K]
v	molar volume [m^3 mol^{-1}]
x_i	mole fraction of component i [-]
Z	compressibility factor [-]

Greek Symbols and Special Characters

γ_i	activity coefficient of component i [-]
Δh_{0i}^{SL}	melting enthalpy of component i [J mol^{-1}]
ε/k	dispersion energy, pure-component parameter of PC-SAFT [K]
$\varepsilon^{A_iB_i}/k$	association energy, pure-component parameter of PC-SAFT [K]
σ	temperature-independent segment diameter, pure-component parameter of PC-SAFT [Å]
φ_i	fugacity coefficient of component i in the mixture [-]
φ_{0i}	fugacity coefficient of pure component i [-]
$\kappa^{A_iB_i}$	association volume, pure-component parameter of PC-SAFT
ν_i	stoichiometric coefficient [-]
μ	chemical potential [J]

Superscripts

assoc	contribution due to associative attraction
disp	contribution due to dispersive attraction
hc	residual contribution of hard-chain system
hs	residual contribution of hard-sphere system
L	liquid phase
res	residual contribution
S	solid phase
SL	melting property

Subscripts

i, j	component i, j
0	pure component, standard state

Appendix

The first section of the appendix provides a summary of the equations to calculate the activity coefficient from the residual Helmholtz energy. The activity coefficient is defined as a quotient of the fugacity coefficient in the mixture φ_i^L to the fugacity coefficient of pure component φ_{0i}^L according to

$$\gamma_i^L \equiv \frac{\varphi_i^L}{\varphi_{0i}^L} \tag{2.A.1}$$

The fugacity coefficient in the mixture is related to the residual chemical potential μ_i^{res}, the Boltzmann's constant k, temperature T, and to the compressibility factor Z according to:

$$RT \ln \varphi_i = \frac{\mu_i^{res}}{kT} - RT \ln Z \tag{2.A.2}$$

The compressibility factor Z is defined as a function of system pressure p, molar volume v, Boltzmann constant k, Avogadro number N_{AV}, and temperature T:

$$Z \equiv \frac{pv}{kN_{AV}T} \tag{2.A.3}$$

The residual chemical potential is the partial derivation of the residual Helmholtz energy A^{res} divided with number of molecules N with respect to the concentration x_i at constant temperature and volume and the concentrations of other components in the system according to:

$$\frac{\mu_i^{res}}{kT} = \frac{A^{res}}{NkT} + Z - 1 + \left(\frac{\partial \frac{A^{res}}{NkT}}{\partial x_i} \right)_{T,v,x_{k \neq i}} - \sum_j \left[x_j \left(\frac{\partial \frac{A^{res}}{NkT}}{\partial x_j} \right)_{T,v,x_{k \neq j}} \right] \tag{2.A.4}$$

With Equations 2.A.2 and 2.A.4, the fugacity coefficient can finally be calculated as:

$$\ln \varphi_i = \frac{A^{res}}{NkT} + \left(\frac{\partial \frac{A^{res}}{NkT}}{\partial x_i} \right)_{T,v,x_{k \neq i}} - \sum_j \left[x_j \left(\frac{\partial \frac{A^{res}}{NkT}}{\partial x_j} \right)_{T,v,x_{k \neq j}} \right] + Z - 1 - \ln Z \tag{2.A.5}$$

The second section provides a summary of the equation of state contributions to the residual Helmholtz energy according to PC-SAFT (Equation 2.10).

Hard-chain Reference Contribution

The Helmholtz energy of the hard-chain reference term is given as

$$\frac{A^{hc}}{NkT} = \bar{m} \cdot \frac{A^{hs}}{N_s kT} - \sum_i x_i(m_i - 1) \cdot \ln g_{ii}^{hs}(d_{ii}) \tag{2.A.6}$$

where x_i is the mole fraction of chains of component i, m_i is the number of segments in a chain and the mean segment number in the mixture is defined as:

$$\bar{m} = \sum_i x_i m_i \tag{2.A.7}$$

The Helmholtz energy for the hard-sphere segments $A^{hs}/N_s kT$ in Equation 2.A.6 is given on a per-segment basis as

$$\frac{A^{hs}}{N_s kT} = \frac{1}{\zeta_0}\left[\frac{3\zeta_1\zeta_2}{(1-\zeta_3)} + \frac{\zeta_2^3}{\zeta_3(1-\zeta_3)^2} + \left(\frac{\zeta_2^3}{\zeta_3^2} - \zeta_0\right)\cdot\ln(1-\zeta_3)\right] \tag{2.A.8}$$

where N_s is related to the number of hard-spheres and the radial pair distribution function for the hard-sphere fluid is given by

$$g_{ij}^{hs}(d_{ij}) = \frac{1}{(1-\zeta_3)} + \left(\frac{d_i d_j}{d_i + d_j}\right)\frac{3\zeta_2}{(1-\zeta_3)^2} + \left(\frac{d_i d_j}{d_i + d_j}\right)^2 \frac{2\zeta_2^2}{(1-\zeta_3)^3} \tag{2.A.9}$$

and ζ_n is defined as:

$$\zeta_n = \frac{\pi}{6}\cdot\rho\sum_i x_i m_i d_i^n \quad n = \{0,1,2,3\} \tag{2.A.10}$$

The temperature-dependent segment diameter is obtained as

$$d_i = \sigma_i\left(1 - 0.12\cdot\exp\left(-3\cdot\frac{\varepsilon_i}{kT}\right)\right) \tag{2.A.11}$$

where σ_i is the temperature-independent segment diameter and ε_i is the depth of the pair-potential.

Dispersion Contribution

The dispersion contribution to the Helmholtz energy is given by

$$\frac{A^{disp}}{NkT} = -2\pi\rho\cdot I_1(\eta,\bar{m})\cdot\sum_i\sum_j x_i x_j m_i m_j\left(\frac{\varepsilon_{ij}}{kT}\right)\sigma_{ij}^3$$

$$-\pi\rho\cdot\bar{m}\cdot C_1\cdot I_2(\eta,\bar{m})\cdot\sum_i\sum_j x_i x_j m_i m_j\left(\frac{\varepsilon_{ij}}{kT}\right)^2\sigma_{ij}^3 \tag{2.A.12}$$

with

$$C_1 = \left(1 + Z^{hc} + \rho\frac{\partial Z^{hc}}{\partial\rho}\right)^{-1}$$

$$= \left(1 + \bar{m}\frac{8\eta - 2\eta^2}{(1-\eta)^4} + (1-\bar{m})\frac{20\eta - 27\eta^2 + 12\eta^3 - 2\eta^4}{[(1-\eta)(2-\eta)]^2}\right)^{-1} \tag{2.A.13}$$

with η is the packing fraction which equals to ζ_3 in Equation 2.A.10.

The power series I_1 and I_2 depend only on the packing fraction and segment number according to:

$$I_1(\eta,\bar{m}) = \sum_{i=0}^{6} a_i(\bar{m}) \cdot \eta^i \qquad (2.A.14)$$

$$I_2(\eta,\bar{m}) = \sum_{i=0}^{6} b_i(\bar{m}) \cdot \eta^i \qquad (2.A.15)$$

where the coefficients $a_i(m)$ and $b_i(m)$ are functions of the segment number:

$$a_i(\bar{m}) = a_{0i} + \frac{\bar{m}-1}{\bar{m}} a_{1i} + \frac{\bar{m}-1}{\bar{m}} \frac{\bar{m}-2}{\bar{m}} a_{2i} \qquad (2.A.16)$$

$$b_i(\bar{m}) = b_{0i} + \frac{\bar{m}-1}{\bar{m}} b_{1i} + \frac{\bar{m}-1}{\bar{m}} \frac{\bar{m}-2}{\bar{m}} b_{2i} \qquad (2.A.17)$$

The universal model constants in Equations 2.A.11 and 2.A.12 are given in the work of Gross and Sadowski [9].

Association Contribution

The association contribution to the Helmholtz energy is given as:

$$\frac{A^{assoc}}{NkT} = \sum_i x_i \sum_{A_i=1}^{nsite} \left(\ln X^{A_i} - \frac{X^{A_i}}{2} + \frac{1}{2} \right) \qquad (2.A.18)$$

It is important to note that the summation runs over all association sites of the molecule i where X^{A_i} is the fraction of the free molecules i that are not bonded at the association site A:

$$X^{A_i} = \left(1 + \rho \cdot \sum_j x_j \sum_{B_j}^{nsites} X^{B_j} \cdot \Delta^{A_i B_j} \right)^{-1} \qquad (2.A.19)$$

with

$$\Delta^{A_i B_j} = g_{ij}^{hs} \cdot (d_{ij}) \cdot \kappa^{A_i B_j} \cdot \sigma_{ij}^3 \cdot \left(\exp\left(\frac{\varepsilon^{A_i B_j}}{kT} \right) - 1 \right) \qquad (2.A.20)$$

where $g_{ij}^{hs} \cdot (d_{ij})$ is the pair distribution function of hard spheres given in Equation 2.A.9.

References

1 Casas, L., et al. (2008) *J. Supercrit. Fluids*, **45** (1), 37.

2 Mukhopadhyay, M. (2009) *J. Chem. Technol. Biotechnol.*, **84** (1), 6.

3 Rytting, E., et al. (2004) *Pharm. Res.*, **21** (2), 237.

4 Huuskonen, J., Livingstone, D.J. and Manallack, D.T. (2008) *SAR*

QSAR Environ. Res., **19** (3–4), 191.
5 Klopman, G. and Zhu, H. (2001) *J. Chem. Inf. Comput. Sci.*, **41** (2), 439.
6 Delaney, J.S. (2004) *J. Chem. Inf. Comput. Sci.*, **44** (3), 1000.
7 Chen, C.C. and Crafts, P.A. (2006) *Ind. Eng. Chem. Res.*, **45** (13), 4816.
8 Klamt, A., et al. (2002) *J. Comput. Chem.*, **23** (2), 275.
9 Gross, J. and Sadowski, G. (2001) *Ind. Eng. Chem. Res.*, **40** (4), 1244.
10 Gupta, R.B. and Heidemann, R.A. (1990) *AIChE J.*, **36** (3), 333.
11 Chapman, W.G., Jackson, G. and Gubbins, K.E. (1988) *Mol. Phys.*, **65** (5), 1057.
12 Wertheim, M.S. (1984) *J. Stat. Phys.*, **35** (1–2), 19.
13 Wertheim, M.S. (1984) *J. Stat. Phys.*, **35** (1–2), 35.
14 Wertheim, M.S. (1986) *J. Chem. Phys.*, **85** (5), 2929.
15 Barker, J.A. and Henderson, D. (1967) *J. Chem. Phys.*, **47** (8), 2856.
16 Barker, J.A. and Henderson, D. (1967) *J. Chem. Phys.*, **47** (11), 4714.
17 Kleiner, M. and Sadowski, G. (2007) *J. Phys. Chem. C*, **111** (43), 15544.
18 Chapman, W.G., Gubbins, K.E., Jackson, G. and Radosz, M. (1990) *Ind. Eng. Chem. Res.*, **29** (8), 1709.
19 Wolbach, J.P. and Sandler, I.S. (1998) *Ind. Eng. Chem. Res.*, **37** (8), 2917.
20 Fuchs, D., Fischer, J., Tumakaka, F. and Sadowski, G. (2006) *Ind. Eng. Chem. Res.*, **45** (19), 6578.
21 Ruether, F. and Sadowski, G. (2009) *J. Pharm. Sci.*, **98**, 4205.
22 Abildskov, J. (2005) *Solubility and Related Properties of Large Complex Chemicals, Part 2: Organic Solutes Ranging from C2 to C41*, DECHEMA, Frankfurt am Main, Germany.
23 Marrero, J. and Abildskov, J. (2003) *Solubility and Related Properties of Large Complex Chemicals, Part 1: Organic Solutes Ranging from C4 to C40*, DECHEMA, Frankfurt am Main, Germany.
24 Nielsen, T.L., et al. (2001) *J. Chem. Eng. Data*, **46** (5), 1041.
25 Marrero, J. and Gani, R. (2001) *Fluid Phase Equilibria*, **183–184**, 183.
26 Ruchelman, M.W. (1969) *Anal. Biochem.*, **19**, 98.
27 Ruchelman, M.W. and Howe, C.D. (1969) *J. Chromatogr. Sci.*, **7** (6), 340.
28 Bar, L.K., Garti, N., Sarg, S. and Bar, R. (1984) *J. Chem. Eng. Data*, **29** (4), 440.
29 Hansen, B.N., et al. (2001) *J. Chem. Eng. Data*, **46** (5), 1054.
30 Sovova, H., Stateva, R.P. and Galushko, A.A. (2001) *J. Supercrit. Fluids*, **21** (3), 195.
31 Johannsen, M. and Brunner, G. (1997) *J. Chem. Eng. Data*, **42** (1), 106.
32 Craft, N.E. and Soares, J.H. (1992) *J. Agric. Food Chem.*, **40** (3), 431.
33 Treszczanowicz, T., Kasprzycka-Guttman, T. and Treszczanowicz, A.J. (2003) *J. Chem. Eng. Data*, **48** (6), 1517.
34 Kraska, T., Leonhard, K.O., Tuma, D. and Schneider, G.M. (2002) *J. Supercrit. Fluids*, **23** (3), 209.
35 Mendes, R.L., Nobre, B.P., Coelho, J.P. and Palavra, A.F. (1999) *J. Supercrit. Fluids*, **16** (2), 99.
36 Granberg, R.A. and Rasmuson, A.C. (1999) *J. Chem. Eng. Data*, **44** (6), 1391.
37 Lorenz, H., Sapoundjiev, D. and Seidel-Morgenstern, A. (2002) *J. Chem. Eng. Data*, **47** (5), 1280.

3
Alternative Solvents in Plant Extraction
Volkmar Jordan and Ulrich Müller

3.1
Introduction

The extraction of active agents from plants most prevalently involves solvents from organic sources and alcohol–water mixtures. Substances characterized as such are described in an overview [1] with the focus on the solubility of triterpenic acids in various solvents. For special applications, like the production of caffeine, the use of supercritical carbon dioxide in the laboratory and production scale setting is advantageous, mainly because carbon dioxide can be easily separated from the extracts, thus leaving no traces of solvent in the final product. However, this benefit is offset by high energy and capital costs accumulating from the use of high pressure technology. Hence, there is high motivation to find alternative solvents which are nontoxic and relevant the food and cosmetic industry. In addition to supercritical carbon dioxide, water, polyethylene glycol, fluorinated phases and ionic liquids are discussed [2]. Furthermore, systems based on surfactants are well known in the area of biotechnology for the extraction of protein from cells. Several papers and patents (see Table 3.4 in Section 3.3.6) demonstrate applications of the micellar and the cloud point extraction.

In this chapter, ionic liquids and surfactants are of central interest. The former class of solvents is already used for liquid–liquid extraction [3–7] and its potential for the recovery of organic and bio- molecules has already been described [6]. Several examples show that ionic liquids are applicable to the removal of biomolecules from plant material or for further enrichment of those components extracted with conventional solvents [8–11]. The newest developments are based on the creation of ionic liquids with structural elements like hydroxyl-, carboxyl- or amino groups on the cation. This new class of ionic liquids of the third generation [2] offers task-specific solvents.

3.2
Ionic Liquids in the Extraction of Natural Compounds from Plant and Fungi

Ionic liquids play an increasingly important role in separation technology [12]. For both liquid–liquid and solid–liquid extraction the number of publications strongly increased during the last decade. Table 3.1 shows the reported examples with ionic liquids (ILs) used in solid–liquid extraction. Furthermore two additional examples where ILs are used in liquid–liquid extraction and solid-phase extraction processes

Table 3.1 Examples for the extraction of biomolecules with ionic liquids.

Extracted molecule	Ionic liquid	Carrier matrix	Application	Reference
Lignin	[EMIM][CH3COO]	Maple wood	Pre-treatment and recovery	[13]
Lignin fragments	[EMIM][ABS]	Sugar cane begasse	Recovery	[14]
Artemisinin	[DMEA]oct [BMOEA]bst	*Artemesia annua*	Recovery	[8]
Akaloids (codeine, morphine, ...)	Alkolyl-, alkoxyalkyl-, aminoalkyl-substituted ammonium salt	Plant and fungi	Recovery	[9]
Phenolic alkaloids	1-alkyl-3-methylimididazolium derivatives in aqueous solution	*Nelumbo nucifera Gaertn.* (lotus flower)	Analysis	[10]
Trans-resveratrol	[BMIM]Br [BMIM]Cl [BMIM]BF$_4$ in aqueous solution	Root of giant knotweed (*Rhizoma polygoni cuspidate*)	Analysis	[11]
Piperine	1-alkyl-3-imidazolium ILs	White pepper	Analysis	[15]
Gallic acid, ellagic acid, quercetin, trans-resveratrol	[BMIM]Cl, [BMIM]Br, [BMIM][BF$_4$], [EMIM]Br, [HMIM]Br, [BMIM][N(CN)$_2$], [BMIM]$_2$[SO$_4$], [BPy]Cl, [BMIM][H$_2$PO$_4$]	Medicinal plants: *Psidium guajava Smilax china*	Recovery	[16]
Amino acids	[BMIM][PF$_6$], [HMIM][PF$_6$], [HMIM][BF$_4$], [OMIM][BF$_4$]	Aqueous solution	Recovery	[17]
α-tocopherol	[EMIM]glycine	Soybean oil deodorizer distillate	Analysis, solid phase extraction	[18]

are shown. These examples illustrate that ILs can also be used in a second enrichment step to remove the target component from a crude extract produced in a first solid–liquid extraction.

3.2.1
Characteristics of Ionic Liquids

3.2.1.1 Physicochemical Properties

Ionic liquids can be regarded as molten salts because they consist of cations and anions in nondissociated form. The cation is generally organic, and as such, constitutes the main difference to a normal salt [19]. Ionic liquids have relatively low melting points, normally below 100 °C, and those ILs which are liquid at room temperature (RTILs) are of special interest for separation processes. Figure 3.1 shows some common cation and anion pairs used in ILs and also demonstrates that the chemical properties of ILs may be altered by changing the ion combination as exemplified by hydrophobicity. It is obvious that the miscibility can be altered

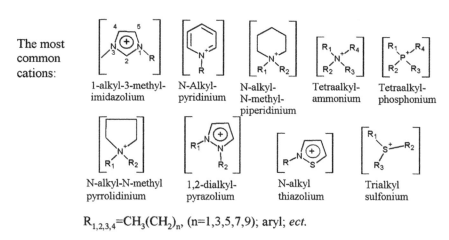

$R_{1,2,3,4} = CH_3(CH_2)_n$, (n=1,3,5,7,9); aryl; *ect.*

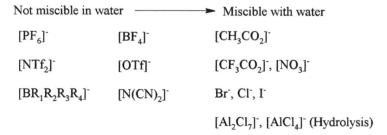

Figure 3.1 Examples of common cation an anion pairs used in the formation of ILs and change of hydrophobicity [2].

Table 3.2 Common abbreviations for cations and anions of ionic liquids.

Abbreviation	Full name of anion or cation
[MMIM]$^+$	1-methyl-3-methylimidazolonium cation
[EMIM]$^+$	1-ethyl-3-methylimidazolonium cation
[HMIM]$^+$	1-hexyl-3-methylimidazolonium cation
[OMIM]$^+$	1-octyl-3-methylimidazolonium cation
[BMOEA]$^+$	bis 2-methoxyethylammonium cation
[DMEA]$^+$	N,N-dimethylethanolammonium cation
[NTf$_2$]$^-$	bis fluoromethylsulfonyl amid anion
[OTf]$^-$	trifluoromethane sulfonate anion
Oct	octanoate anion
Bst	bis trifluoromethysolfonyl imide anion
BF$_4^-$	tetrafluoroborate
PF$_6^-$	hexafluorophosphate

from miscible to completely nonmiscible with water. Abbreviations for the long names of the cations and anions important for the ILs discussed in this contribution are explained in Table 3.2.

The liquid range of RTILs can be much greater than a common organic solvent or water and has been reported to vary almost 400 °C (from −90 °C to 300 °C) [19]. The vapor pressure of ILs is very low or negligible, and therefore, the upper limit of the liquid range is not given by evaporation. The use of ILs is limited by decomposition. The upper temperature limit lies between 250 and 400 °C and is dependent on the ion pair used [12, 19]. The thermal decomposition temperature of the ILs depends on the coordinating ability of the anion. Based on data from thermogravimetric analysis onset measurements by different authors given in [19], it can be seen that the stability of ILs with 1-alkyl-3-methylimidazolium cations follows the order Cl$^-$ < [BF$_4$]$^-$ < [PF$_6$]$^-$ < [NTf$_2$]$^-$.

Based on historical development and progress, the distinction between first, second and third generation ionic liquids [2] is made:

1st generation: These ILs were developed for electrochemical applications and are highly sensitive to hydrolysis.
2nd generation: First-generation ILs that carry a modified anion to become immiscible in water.
3rd generation: Task-specific ILs that were created by insertion of special functional groups to the cation.

Viscosity and density are important parameters that affect the choice of an extraction solvent because they influence the separability from the second phase. A high viscosity and low density difference will lead to a more complex solid–liquid separation after extraction. Furthermore, the pressure drop in the percolation process is negatively influenced by increased viscosity while the mass transfer rate also decreases. Thus, extraction kinetics is adversely affected.

To date, researchers have treated ILs as Newtonian fluids, and so far no data have been published to indicate that there are non-Newtonian ILs [19]. Ionic liquids are more viscous than most common solvents used in the extraction process because organic solvents and water show viscosities in the range of 1 cP while collected data in [19] show that the IL viscosity varies between 10 and more than 500 cP. Even in low concentrations, impurities are of significant relevance because they have a strong effect on viscosity [20, 21]. Furthermore, it has been reported that the temperature has a great impact on the viscosity of ILs as a 27% change in viscosity results from a temperature variation of 5 °C in the case of some special ILs [19].

Impurities have little influence on the density of ILs, which varies from $1120\,kg\,m^{-3}$ to $2400\,kg\,m^{-3}$. For example, 20% of water in [BMIM][BF4] results in only a 4% decrease in density. A temperature change does not have much effect on the density either [19].

Water is omnipresent in ILs if they are not handled in a completely inert manner or in the absence of any special drying procedures. The uptake of water from the air is exhibited by hydrophobic ILs to such an extent that the resulting saturation is 1.4 percent by mass. For more hydrophilic ILs the water uptake can be much higher [19]. As aforementioned, the water content has an influence on the viscosity and the capacity and selectivity during the extraction processes; and therefore, identification of the exact concentration of water in ILs to be tested during the process development is of great importance.

Ionic liquids differ from polar solvents like water or methanol and by combination of the anion and cation it is possible to design highly selective solvents. Hydrocarbons are poorly soluble in ILs but not insoluble. It is reported that alkanes are insoluble in all ILs [22] while alkenes show a low solubility increasing with increased alkyl chain substitution in the IL.

Ionic liquids generally show partitioning properties similar to those of dipolar aprotic solvents [19]. The partition coefficient of an IL–water system correlates with that from an octanol–water system. Complex organic molecules like cyclodextrins, glycolipids, antibiotics [19], artemisinin [8] and alkaloids [9] can be dissolved in ILs. Aromatic compounds are considerably more soluble than aliphatics [19]. A ratio of up to 1 : 1 between IL and solute has been demonstrated with special ILs for benzene. The partitioning of amino acids showed that the partition coefficient changes with the pH and that amino acids like tryptophan, which contains an aromatic group, exhibit higher partition coefficients than aliphatic amino acids [18]. Charged and ionizable solutes show pH-dependent partitioning in IL–water systems [5]. Solute ionization can change the solubility and partitioning by several orders of magnitude (see Figures 3.2 and 3.3).

From empirical observation, ILs tend to be immiscible with nonpolar solvents. Thus, ILs can be washed or brought into contact with diethyl ether or hexane to extract nonpolar products. With increasing solvent polarity, esters like ethyl acetate exhibit variable solubility with ILs, depending on their nature. Polar solvents (including chloroform, acetonitrile, and methanol) appear to be totally miscible with all ILs (an exception is tetrachloroaluminate ILs and the like, which is capable

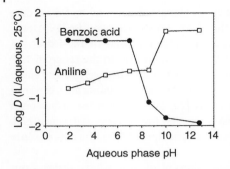

Figure 3.2 Distribution ratios for aniline (pK_b = 9.42) and benzoic acid (pK_a = 4.19) in [BMIM][PF$_6$]–aqueous systems as a function of aqueous phase (Reproduced with permission from[19]. Copyright © (2008) Wiley-VCH Verlag GmbH.).

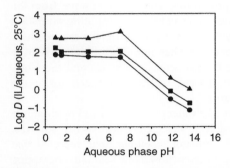

Figure 3.3 pH switchable partitioning of the ionic dye thymol blue in [BMIM][PF$_6$], [OMIM][PF$_6$], [HMIM][PF$_6$]–aqueous biphasic systems. (Reproduced with permission from[19]. Copyright © (2008) Wiley-VCH Verlag GmbH.)

of reacting). Among notable exceptions, [EMIM]Cl and [BMIM]Cl are insoluble in dry acetone[19]. It was shown that hydrophobic hexafluorophosphate ILs can be designed totally miscible in water [19]. By the addition of ethanol the ternary system ethanol–IL–water shows a large co-miscibility region of the three components [23]. In fact, ethanol forms biphasic layers with [BMIM][PF$_6$], [HMIM][PF$_6$] and [OMIM][PF$_6$]. The degree of miscibility is dependent on temperature and water content of the IL. In all cases, an increase in the IL water content leads to higher ethanol solubility.

3.2.1.2 Environmental and Safety Aspects

Ionic liquids are often proposed as green solvents due to their low or negligible volatility which is advantageous in that processes can be designed without the emission of volatile organic compounds (VOCs). On the other hand, many ILs are water-soluble and can enter the environment via this path [19] or they are detectable in the products in small concentrations. Hence, it is necessary to assess this new class of solvents with respect to toxicity, ecotoxicity and biodegradability. A good overview is given in [19, 24], showing that the toxic potential of ILs is clearly correlated to the chain length of the alkyl chains on the cation. Longer chain length leads to increased toxicity.

With respect to applications in the plant extraction process, biodegradability plays an important role (see Section 3.2.2.2). It is reported that an octyl sulfate anion is more biodegradable than other commonly used anions. The introduction of an ester group in the side chain of the 1,3-dialkylimidazolium cation leads to biodegradation values very close to passing a closed bottle test [25]. ILs are deemed "readily biodegradable" in the presence of a pyridinium cation with an ester side chain and an octyl sulfate anion [26].

3.2.2
Application of Ionic Liquids in Plant Extraction

3.2.2.1 Application of ILs in Extraction

A variety of different applications have been published for liquid–liquid extractions, however only a few publications show the potential of ILs in solid–liquid extraction [12]. Some authors show the use of such solvents in small scale applications mainly used for sample preparation in chromatography [10, 11, 13]. These publications already indicate the potential of ILs for the extraction of organic substances from plants and fungi, namely alkaloids, phenolics, and tocopherol. Another example is the removal of the phenolic macromolecule lignin from sugar cane plant waste [15] at elevated temperatures with an extraction yield of 93% reported. The company Bioniqs Ltd (UK), as a producer of ILs, is very active in the field of solid–liquid extraction from plant material as shown by several publications [8, 9] which include the extraction of artemisinin and several alkaloids with ILs. The examples morphine, caffeine and opiates (morphine, codeine, thebaine) show high capacities compared with normal organic solvents [10]. For morphine extraction with 1-butyl-3-methylimidazolium hexafluorophosphate [BMIM][PF$_6$] a solubility of $78\,\text{g}\,\text{L}^{-1}$ is reported while the maximum concentration in chloroform is only $3.3\,\text{g}\,\text{L}^{-1}$.

3.2.2.2 Removal of Target Substance from Extract and Separation of Solvent from Spent Biomass

Ionic liquids have a very low or negligible vapor pressure. For only a few special ILs of the 1-n-alkyl-3-methylimidazolonium bis(trifluoromethanesulfonylimide) type is it reported that they could be vacuum distilled at 300 °C and 0.1 mbar [27]. In this they differ from normal organic solvents, which have a high volatility. If we consider the example of oil extraction with hexane, it obvious that the separation of the solvent from spent biomass and from the extracted oil can easily be done by evaporation. If ILs are used one has to find other ways to recover the solvent from the biomass and to remove the solvent from the extracted target components.

Lapkin et al. [8] point out that solvent recovery from the plant after the extraction is still a problem to be solved. Two principles seem possible: a removal of the ILs can be avoided if cheap solvents which are biodegradable are used or if the biomass is energetically used and incinerated after extraction. If the ILs are highly valuable and the solvent loss is unacceptably high, the biomass after extraction can be treated with a washing liquid with high volatility. The washing liquid can be separated from the IL by evaporation.

For the removal of the target substance several ways are proposed [5, 8, 12]. Lapkin et al. [8] use a water soluble IL. By addition of water to the extract the solubility of the extracted substance is reduced and the product is separated from the IL by crystallization or precipitation. More details on this method are discussed in Section 3.2.2.3 with the target substance artemisinin. Lignin, which is dissolved in a mixture of an IL with an [EMIM] cation and alkylbenzenesulfonate (ABS) anion and some percentage of water, can be recovered by lowering the pH to 2 [15]. Back extraction into a water phase by changing the pH can be possible and supercritical CO_2 can be used, too [5, 12]. Furthermore, organic liquids forming two-phase systems with the IL can be used for the back extraction of neutral, thermally unstable substances [12] and volatile ones can be removed from the IL via distillation.

3.2.2.3 Example 1: Extraction of Artemisinin

The best developed example is the extraction of artemisinin (see Figure 3.4) from *Artemesia annua*. Its extraction with an ionic liquid is compared with the use of other solvents like hexane, ethanol, supercritical CO_2 and hydrofluorocarbons (HFC-134a) [8, 9].

The solubility of artemisinin in the ionic liquids *N,N*-dimethylethanolammonium octanoate (DMEAoct) and bis 2-methoxyethylammonium bis(trifluoromethysulfonyl imide) (BMOEA bst) is in the range of ethyl acetate and much higher than in either hexane and ethanol (see Figure 3.5) [8]. This data (see Table 3.3) shows that these ionic liquids can be very effective solvents in extraction processes.

The IL-based extraction process is illustrated in Figure 3.6. It involves the usage of dry biomass suspended in the solvent at 25 °C wherein the solvent to biomass ratio lies between 6.3 : 1 and 9 : 1 (w/w). Compared with the solubility data in

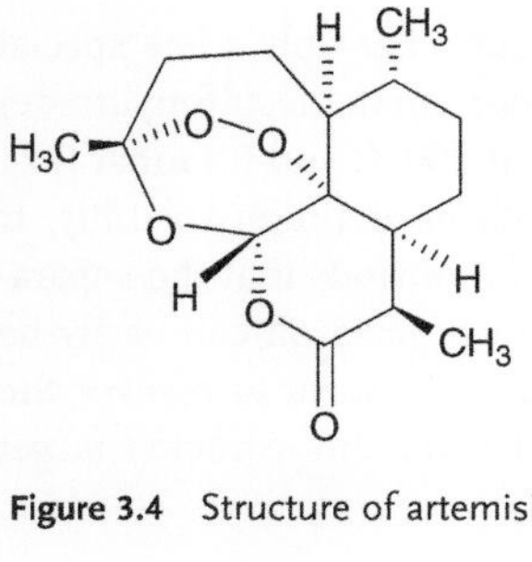

Figure 3.4 Structure of artemisinin.

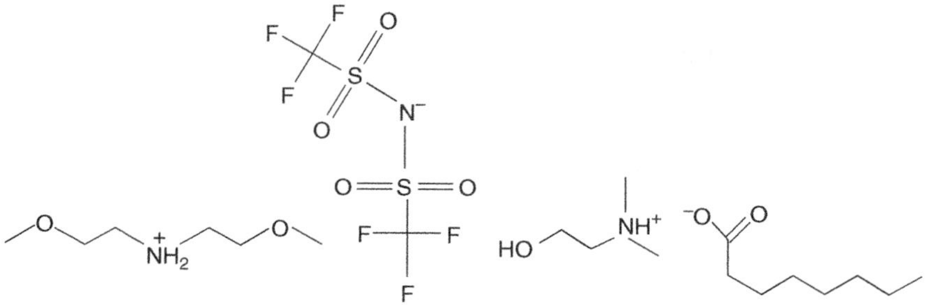

Figure 3.5 Structure of DMEA oct (left) and BMOEA bst (right) [8].

Table 3.3 Solubility data of artemisinin in ionic liquids [8].

Ionic Liquid	Solubility in g L^{-1}
DMEA oct	82
BMOEA bst	110

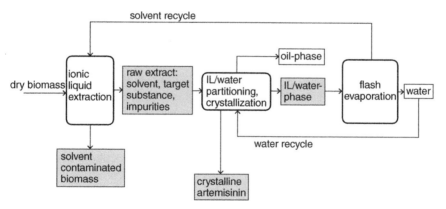

Figure 3.6 Schematic diagram of the artemisinin extraction with ionic liquids (Reproduced with permission from [8]. Copyright © (20067) American Chemical Society.).

Table 3.3 the concentrations in the extract are with 0.08 to 0.6 wt% much smaller. The extraction is performed at ambient temperature over a time period of 30 min, after which water is added so that the separation of the oil fraction and crystallization of the target product artemisinin occurs simultaneously. The crystallization allows a recovery of 82% of the extracted artemisinin. The crystals have a purity of 95% and are essentially free of solvent [8]. The authors compare the IL-based

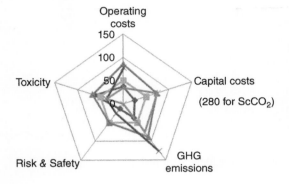

Figure 3.7 Multi-objective comparison of extraction technologies with respect to operating costs, capital costs, greenhouse gas (GHG) emissions, risk&safety and toxicity (Reproduced with permission from [8]. Copyright © (20067) American Chemical Society.).

extraction to extractions with hexane, ethanol, supercritical CO_2 and HFC-134a. The result of this comparison is shown in Figure 3.7. Obviously ILs exhibit a high potential to substitute hexane or ethanol in the artemisinin extraction and compared with scCO_2 they show a high cost advantage due to the capital costs of high pressure processes. The traditional extraction and purification process is described in [28]. This process uses three different solvents in the stages extraction, enrichment and purification. Furthermore charcoal is needed and the silica gel used for the removal of waxes by column chromatography cannot be completely regenerated. This complex multistage procedure, necessary to remove the high amount of numerous coextracted compounds, shows the incentive for the development of new processes based on more selective solvents used in the solid–liquid extraction.

3.2.2.4 Example 2: Extraction of Lignin

The extraction of lignin as a pre-treatment for improved enzymatic hydrolysis of cellulose [14] or for the use of waste biomass [15] is easily possible with the help of ILs. Using an optimized combination of cation and anion makes it possible to find ILs with low cellulose and high lignin solubility. [Cl]⁻-containing ILs show a high cellulose solubility while [MMIM][MeSO$_4$], [BMIM][CF$_3$SO$_3$] or [EMIM][CH$_3$COO] show high lignin solubility. Lignin solubilities of 300 to 500 g kg^{-1} are reported. Experiments were conducted with [EMIM][CH$_3$COO] at 80 °C due to its advantages for subsequent enzymatic hydrolysis. The IL may be reusable if the remaining solid is washed with water and the IL dried by evaporating all of the washing water [14]. Lignin can be recovered by a method in which the pH is

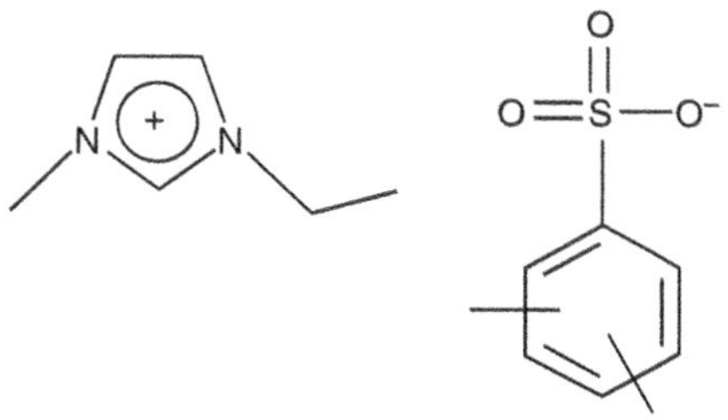

Figure 3.8 The ionic liquid 1-ethyl-3-methylimidazolium xylenesulfonate.

changed to the acidic range [15]. For the extraction of lignin from begasse (a waste biomass of the sugar production based on sugar cane) tailor-made ILs with an [EMIM] cation and a xylenesulfonate anion (see Figure 3.8) can be used in a reactive extraction at elevated temperatures (190 °C) The initial water concentration of the mixture is 30% because a specific breakdown in the size of the lignin macromolecule is desirable. During the treatment of the begasse the water concentration is reduced to 2% due to the hydrolytic fragmentation of the lignin. The process exhibits yields above 93%, but the lignin differs from its commercially available counterpart because of ABS adducts, which result from the reaction of the [ABS] anion with the carbonium ion in the lignin fragments [15].

3.3
Surfactants and Aqueous Two-Phase Systems in Plant Extraction

In this chapter an overview is presented of surfactants which form micellar systems in aqueous solution. Surfactants are amphiphilic compounds consisting of a hydrophilic part and a lipophilic part. The hydrophilic part has a polar group with an affinity for polar solvents particularly water whereas the lipophilic part has a nonpolar group with an affinity for nonpolar substances like organic solvents. Due to the amphiphilic structure surfactants are able to reduce the surface tension of a liquid, allowing easier spreading, and reduce the interfacial tension between two liquids. Therefore, they are soluble in both organic solvents and water. In Figure 3.9 the molecular structure of a phosphatide is shown as an example of such amphiphilic molecules and additionally the generally used pictogram where a circle stands for the hydrophilic and a line for the lipophilic part of this molecule.

Surfactants can be classified by a variety of factors [29]. The commonly used chemical classification, concerning the charge of the polar hydrophilic group, divides surfactants into two groups, the "ionic" and "nonionic" surfactants. The ionic surfactants are classified as anionics, cationics and amphoterics, often referred to as zwitterionics (see Figure 3.10).

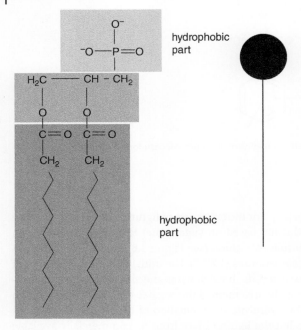

Figyre 3.9 Molecular structure of a phosphatide surfactant and the pictogram as a symbol for the amphiphilic structure.

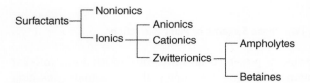

Figyre 3.10 Classification of surfactants.

Anionic surfactants
The anionic surfactants contain a nonpolar lipophilic and an anionic polar hydrophilic group. Carboxylate, sulfate, sulfonate and phosphate are the polar groups found in anionic surfactants. Alkali alkanoates, or soaps, are the best known anionic surfactants. The ionized carboxyl group provides the negative charge.

Cationic surfactants
The cationic surfactants contain a polar, hydrophilic, anionic group and a nonpolar lipophilic group. The lipophilic part of the molecule contains hydro-carbons while the hydrophilic integrates a nitrogen group and is positively charged. Chloride (Cl^-), bromide or methyl sulfate [CH_3O^-, SO^{3-}) act as the counter-ion. Since

these emulsifiers are derived from quaternary ammonium salts they are often referred to as quats. An important property of these emulsifiers is their ability to cover negatively charged surfaces. The hydrophobizing of surfaces can compensate for electrostatic static charges.

Amphoteric surfactants

Amphoteric surfactants can be sectioned in two groups: "ampholytes" and "betaines". Their function depends on the pH value of the aqueous solution. They can form both cationic and anionic structure. Lecithines are the best known examples of the group of betaines.

Nonionic surfactants

Nonionic surfactants do not form ions. Thus, they do not dissociate in aqueous solution. Most of them are lipophilic hydrocarbon compounds which have a hydrophilic polyoxyethylene group (ethylene oxide/oxyethylene units). The water solubility depends on the number of ethylene oxide units. Nonionic surfactants have either a polyether or a polyhydroxyl unit as the polar group. In the vast majority, the polar group is a polyether consisting of oxyethylene units, formed by the polymerization of ethylene oxide. The typical number of oxyethylene units in the polar chain reaches from five to ten.

3.3.1
Characteristics of Surfactant–Water Mixtures

Dependent on the total polarity of the molecule and controlled by the character of the hydrophilic part, surfactants in low concentration are dissolved in water (see Figure 3.11 part 1). The surfactants tend to be adsorbed at surfaces and phase boundaries like solution–air–surface and interfacial surfaces in the case of existence of additional lipophilic droplets. The surfactant molecules spread over

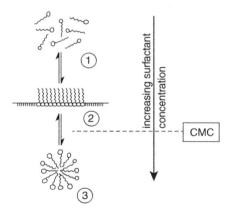

Figure 3.11 Formation of spherical micelles (point 3) from dissolved (1) and surface-adsorbed (2) surfactants by increasing the surfactant concentration from 1 to 3.

the surface while the polar head group is directed to the water and the nonpolar tail to the hydrophobic or gas phase (see Figure 3.11, part 2). At higher concentration further dissolution is not possible and the availability of free surface is reduced. The surfactants start to form micelles (see Figure 3.11, part 3). This phenomenon can be interpreted as precipitation of the surfactant aggregates on the nanoscale. The concentration of surfactant which is the starting point of micelle formation is called critical micelle concentration (CMC). The CMC is strongly dependent on the structure of the surfactant. The micelles which are formed above CMC are the globular or spherical micelles.

Surfactant molecules in micelles are in a dynamic equilibrium with the individual surfactant molecules in solution and at the surface. Therefore, micelles are not rigid structures but are permanently changing their size and shape. Micelles of nonionic surfactants contain 100–1000 molecules, whereas micelles of ionic surfactants are much smaller (50–100 molecules). As mentioned above every surfactant has a specific concentration at which it starts to form micelles. Ionic surfactants are charged and their molecules repel each other. Therefore, they do not start to form micelles until they have reached a relatively high concentration. Nonionic surfactants do not carry a charge. Their critical micelle concentration is appreciably lower. The CMC is an important physical constant for the characterization of surfactants. Many properties of surfactant solutions such as the emulsifying performance do not develop fully until concentrations above the CMC are achieved. A review about the basics of micelle formation and the CMC is given by Texter [29]. The CMC can be determined by measuring the surface tension of a surfactant solution as a function of the surfactant concentration. Above the CMC the surface tension remains constant and is independent on the surfactant concentration. A further increase of the surfactant concentration leads to an association of the micelles. In this case a gelatinous paste with a high viscosity is obtained. At very high surfactant concentrations the solutions become thinner again. So-called mesophases, also referred to as co-azervate or liquid crystal phases, appear. The CMC is defined by the Krafft point which is the temperature where solubility of a surfactant rises sharply and the mixture of surfactant and water becomes liquid. Above the Krafft point the CMC is nearly independent on temperature. The principal context is summarized in Figure 3.12.

If the concentration is much higher than the CMC, the surfactant molecules form aggregates with the characteristics of a liquid as a crystallized body. Because of the quasi-crystallized orders such "mesophases" are also known as liquid crystals. A distinction is drawn between thermotropic and the lyotropic mesomorphism.

Thermotropic mesophases are those that occur in a certain temperature range. A characteristic of such mesophases is the existence of two-phase transition points. At the higher one the thermotropic phase is transformed into a conventional isotropic liquid phase. At the lower one, most thermotropic mesophases will be transformed into a turbid liquid anisotropic crystal. Many thermotropic mesophases exhibit a variety of phases as temperature is changed and the transformations are reversible by heating/cooling.

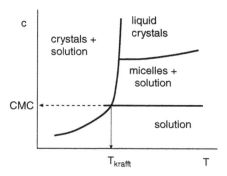

Figure 3.12 Temperature dependence of the critical micelle concentration (CMC) and the role of Krafft point (CE: concentration of surfactant; T: temperature); according to [30].

Amphiphilic molecules form lyotropic mesophases when a solvent, usually water, is added. The process of self-assembly is driven by the hydrophobic effect. In lyotropic mesophases individual aggregates can still be recognized and distinguished by their different optical properties. They are arranged so that repulsion between individual particles is minimal. Lyotropic mesomorphism is always associated with swelling that is incorporation of water. The properties of such lyotropic mesophases depend on the structure of the surfactant, the surfactant/water ratio, and the temperature.

Holmberg et al. [31] give a survey on the critical packing parameters of surfactant molecules and preferred aggregate structures caused by geometrical packing. The authors distinguish six different mesophase structures including the spherical micelle and reduce them to the single molecular structure of the surfactant molecule. They define a ratio between the volume v of a molecule and its cross sectional area a (at the hydrophilic part of the molecule) multiplied with the length l of the molecule (a x length l). If $v/(a\ l) = 1$, the molecule has a cylindrical form whereas it is conic with a ratio of $v/(a\ l) < 1/3$. Surfactants with a ratio > 1 have a size-dominating lipophilic part and are used to produce water-in-oil-emulsions. If the $v/(a\ l)$-value is << 1 the molecule is an oil-in-water emulsifier (see Figure 3.13).

Surfactants with a mirror plane structure ($v/(a\ l) = 1$) prefer to form a so-called lamellar mesophase. Nevertheless, the formation of a certain mesophase type is also affected by the surfactant concentration in aqueous solution and other parameters like the temperature or for ionic surfactants the addition of counter-ions. By increasing the surfactant concentration, the spherical micelles usually turn into long rod- or thread-like micelles [31]. They are also designated as a micellar cubic phase. As a result the viscosity of the solution may increase. The length of these micelles can grow to bi-continuous structures where overlapping and tangling of these thread-like micelles is possible. Also branched structures and at least three-dimensional networks, referred to as a sponge phase, may occur. In the following enumeration the main self-assembling structures (mesophase types) which may be important to an application as extraction systems are shown in Figure 3.13:

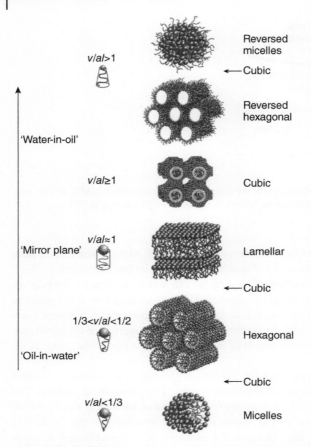

Figure 3.13 Preferred aggregate structures of surfactant-water-systems dependent on the surfactant structure. (Reproduced with permission from [31]. Copyright © (2007) John Wiley & Sons, Ltd, Chichester, UK.)

Hexagonal phase
This phase is composed of infinitely long cylindrical micelles arranged in a hexagonal pattern, with each micelle being surrounded by six others. The radius of the circular cross-section (which may be deformed) is close to that of the surfactant molecule length. [31]

Lamellar phase
This phase is built up of bi-layers of surfactant molecules alternating with water layers. The thickness of the bi-layers is twice the surfactant molecule length. The thickness of the water layer varies dependent on the surfactant type over a wide range. The surfactant bi-layer can range from being stiff and planar to being very flexible and undulating. Lamellar mesophase can be transferred into liposomes being spherical lamellar structures even larger than the spherical

micelle by applying mechanical energy. Liposomes can be used as "containers" for substances with a controlled release which explains their relevance for the pharmaceutical industry. [31]

Bi-continuous phase

Increasing the concentration so-called bi-continuous phases can be obtained. These are structures formed by connecting rod-like micelles (as described above), similar to branched micelles, or bi-layer structures as visualized in Figure 3.13 [31].

Reversed structures

With the exception of the lamellar phase the different structures have a reversed counterpart in which the polar and nonpolar ones take turn. For example, a reverse hexagonal phase is built up of hexagonally packed water cylinders surrounded by the polar heads of surfactant molecules and a continuum of the hydrophobic parts. Reversed (micellar-type) cubic phases and reversed micelles analogously consist of globular water cores surrounded by the surfactant molecules. The radii of these water droplets are typically in the range 20–100 Å. [31]

The phases described are the commonly occurring ones, but in addition there are other mesophases of less importance. Some of these involve discrete aggregates of different shapes and different types of mutual organization and some are similar to the hexagonal phase but with alternative arrangements of the cylinders or noncircular cross-sections of the aggregates. More details can be found in Holmberg et al. [31].

For the application of mesophases as extraction systems of phyto-substances a lot of demands are to be fulfilled, for example, phase separation from the coexisting aqueous phase, selective enrichment of desired substance in the mesophase, and a strategy to recover the target substance from the mesophase. The lamellar mesophase is of high relevance for the extraction of natural compounds. In Figure 3.14 the phase diagram (temperature vs. water concentration) of lecithin and water is shown. The lamellar mesophase (lam.) dominates, and already appears in small lecithin concentrations and coexists with a highly diluted aqueous phase. The cubic (cub.) phase is an interphase as indicated in Figure 3.14.

3.3.2
Behavior of Nonionic Surfactants in Aqueous Solution

Nonionic surfactants have a specific chemo-physical character because their molecular structure consists of the hydrophilic polyethoxylene and the hydrophobic carbon chain. The hydrophilic part of the molecule is surrounded by water molecules and with increasing temperature this hydrate cover is split off. Therefore these nonionic surfactants become more lipophilic and the solubility in water decreases. At a certain temperature a mesophase appears which contains the vast majority of the surfactant precipitated from the solution. This temperature is

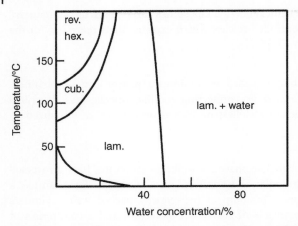

Figure 3.14 Semi quantitative phase diagram of lecithin (from egg yolk) and water, according to [32].

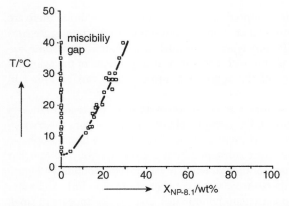

Figure 3.15 Phase diagram (temperature vs. mass concentration) of the nonionic surfactant nonylphenol polyethoxylene with a mean ethoxy-unit of 8.1 (NP-8.1) and molecular structure of NP-8.1, from [33].

named cloud point, regularly measured at a surfactant concentration of 1%. Thus, the phase diagrams exhibit a miscibility gap which becomes bigger with higher temperature. As an example Figure 3.15 shows the phase diagram of the system nonylphenol polyoxyethylene (NP) with water. The mean number of ethoxy units is 8.1 [33].

The miscibility gap disappears at a critical temperature close to the cloud point temperature. For the example shown in Figure 3.15 this temperature is around 4 °C. The cloud point of nonionic surfactant solutions is dependent on the type and concentration of the surfactant. By changing the length of the ethylene oxide

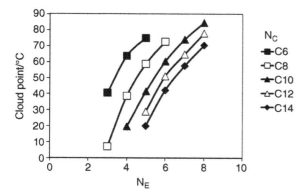

Figure 3.16 Dependence of cloud point on carbon chain length N_C and EO number N_E, for nonionic surfactants, from [34].

chain of polyoxyethylene alkyl ethers the cloud point temperature can be shifted from temperatures below ambient temperature to values close to 90 °C (see Figure 3.16). The cloud point (C in °C) can be empirically described by the equation [34]:

$$C = 220 \log N_E - 5.5 N_C - 55 \tag{3.1}$$

N_E is the number of ethylene oxide units and N_C is the number of carbon atoms in the lipophilic chain. Furthermore, the concentration of certain salts like Na_2SO_4 due to the salting out effect on the ethylene oxide and hydrocarbon chains leads to a decrease of the cloud point. For the surfactant Tergitol 15-S-7 (4wt%), an ethoxylate of a secondary alcohol with hydrocarbon chains having 10 to 14 carbon atoms and an average ethylene oxide number of 7.3, the cloud point can be reduced by addition of 0.4M Na_2SO_4 from 40 to 15 °C [35]. With 0.6 M Na_2SO_4 a cloud point of 5 °C is reached. The cloud point temperature can be adjusted by other substances, too. In [36] the influence of glycerol addition on the cloud point of Triton X114, an octylphenol ethoxylate with an ethylene oxide number of 7 to 8, is depicted.

3.3.3
Micellar Extraction and Cloud Point Extraction

The first step in the removal of the target compounds from the plant with surfactant solutions is a micellar extraction. The spherical micelles formed above CMC are capable of containing hydrophobic substances which are removed from the biomass. In a second step the system is brought into the two-phase region above the cloud point. This leads to an increase in concentration and purity for the extracted hydrophobic substances. Dependent on its nature the target substance is either in the surfactant-rich phase or in the water phase [34, 37]. Because the surfactant-rich mesophase has a volume of 10 to 50% of the total volume a concentration of the target compound by the factor 2 to 10 is feasible. Mainly the lamellar phase is used to absorb hydrophobic and amphiphilic substances. This mesophase can preferably be obtained by using nonionic surfactants. The main problem here is the high cloud point of commercially available surfactants which can lead to an incompatibility with temperature sensitive target compounds. As already discussed in Section 3.3.2 the cloud point can be decreased by using surfactants with a short ethoxy chain or by the addition of salts or other substances. For target compounds which are not temperature sensitive this problem does not occur, by an increase in temperature it is possible to bring the system into the region of an existing miscibility gap.

The first application of such mesophases as an extraction system for biosubstances is reported in the publication of Bordier [38], who described the partition of membrane proteins to the mesophase and the coexisting aqueous phase of Triton surfactants and water. This kind of extractive application of the mesophase is called cloud point extraction.

A lot of work was done in the late 1980s to integrate the mesophase extraction into a fermentative process by a workgroup of Onken, Technical University of Dortmund. The development of extractive fermentation to increase the yield of the antibiotic cycloheximide was successful if nonionic surfactants of the type nonylphenol ethoxylate were used [33]. The phase diagram of this system is shown in Figure 3.15. The partition coefficient of cycloheximide is approximately 10 at 28 °C (fermentation temperature of the bacterium *Streptomyces griseus*). The temperature dependency is shown in Section 3.3.5. Furthermore, Genapol was successfully tested to optimize the production of gibberellic acid. Partition coefficients are summarized in Müller *et al.* [39]. A German patent was submitted in 1988 concerning the extraction of active substances with mesophases [40]. Hinze und Quinia [37] give an excellent review of the publications in the different applications of cloud point extraction until the late 1990s. Since 1990 the number of publication in this field continues to increase. Finally, it must be mentioned that it is also possible to carry out a cloud point extraction with ionic surfactants. Schürholz *et al.* used lecithin mesophases to enrich nicotinic acetylcholine receptors [41].

3.3.4
Reversed Micellar Extraction

Reversed micellar systems are well known for the recovery of biomolecules by liquid–liquid extraction [42–46]. As shown in Table 3.5 reversed micellar extraction can also be applied for the removal of natural compounds from biomass by solid–liquid extraction. Furthermore the micro-emulsions formed by reverse micelle systems can be used for the downstream processing of crude extracts. Figure 3.17 illustrates the principle of the reversed micellar extraction. An aqueous phase is enclosed by a liquid membrane formed by the surfactant molecules and the reverse micelles are dispersed in an organic solvent. Thus, hydrophilic compounds can be absorbed by the aqueous phase.

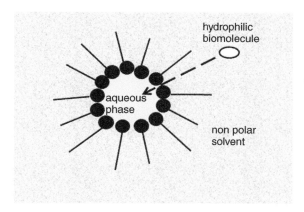

Figure 3.17 The principle of reversed micellar extraction.

The nature of the reverse micelle system depends on the kind of surfactant and solvent to be used. A further important parameter is the surfactant–water-ratio referred to as W_0-value. The optimal W_0-value depends on several aspects. The most important are size and concentration of the target molecule [42]. The selectivity of the extraction can be affected by the W_0-value, too (see Figure 3.21, Section 3.3.6.2).

3.3.5
Equilibrium Partition of Target Substances in Aqueous Surfactant Solutions

Micellar solutions are considered as pseudo two-phase systems. One phase is the aqueous solution with dissolved surfactants as single molecules. The second phase is the micelle which is able to contain hydrophobic substances. The equilibrium partition can be calculated with an extended UNIFAC model. In addition to a combinatorial and a residual contribution a third interfacial term is included. It is calculated based on the Gibbs–Thompson equation [47, 48]. A further improvement is possible if the predictive model COSMO-RS is applied, because ionic components, steric isomers, and inorganic compounds can also be modeled [48].

If surfactant solutions are brought to temperatures above the cloud point, two phases are formed. One is rich in surfactant and the other is at its CMC. Tani et al. [34] give an overview about the theory of partitioning of proteins and present a correlation by Niklas et al. [49] which correlates the partition coefficient K_P with the volume fractions of surfactants in top and bottom phase (Φ_t and Φ_b) depending on the ratio of hydrodynamic radius of protein R_P and cross-sectional radius of the cylindrical micelles R_0.

$$K_p = \exp\left\{-(\Phi_c - \Phi_b)\cdot\left(1+\frac{R_p}{R_0}\right)^2\right\} \quad (3.2)$$

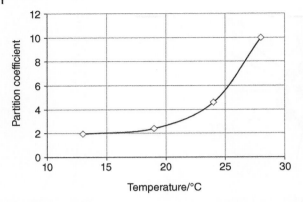

Figure 3.18 Partition coefficient of cycloheximide between mesophase and aqueous phase of the system nonylphenol-8.1-ethoxylate–water (NP-8.1–water). [33]

Hydrophobic and electrostatic interactions are not included in this model. Nevertheless, predictions by Equation 3.2 in the size range 1.9 to 5.2 nm are in good agreement with measured partition coefficients [34]. An enhancement of partitioning is possible by adding affinity or charged ligands. With a hydrophobic affinity ligand an extraction of a hydrophilic compound into the surfactant-rich phase is possible [34]. The partitioning can be affected by many parameters like type of solute, type and concentration of surfactant, temperature, pH, and ionic strength. In particular the influence of salts like Na_2SO_4 used for lowering the cloud point and decreasing the volume of the surfactant-rich phase must be considered. For the model substance naphthalene an increase in salt concentration from 0.4 to 0.6 M Na_2SO_4 leads to a change of the partition coefficient by a factor of 7.4 [35].

Figure 3.18 shows the partition coefficient of the antibiotic cycloheximide between mesophase and aqueous phase of the system nonylphenol-8.1-ethoxylate–water (NP-8.1–water) dependent on temperature [33]. At 28 °C there was a tenfold enrichment of the cycloheximide in the surfactant-rich mesophase where the partition decreases with falling temperature because of the hydrophilic character. Below 13 °C the mesophase and the miscibility gap disappears.

The common parameter to express the hydrophobicity of an organic compound is the octanol–water partition coefficient K_{OW}. Based on this parameter it is possible to describe the partition in a surfactant system above the cloud point. As an example a correlation for polycyclic aromatic hydrocarbons as model substances [35] is presented here:

$$\log_{10} K_p = A \log K_{OW} + B \qquad (3.3)$$

K_p is the partition coefficient in the micelle water system. K_{OW} is the partition coefficient in an octanol–water system. A and B are constants. They show that high preconcentrations and recoveries close to 100% are possible above the cloud point resulting in a partition between water and surfactant phase. But this is only the second step after the initial removal of the target component from the biomass in a first solid–liquid extraction step. For the partition between micellar solution and solid biomass (ground plant parts or fungi) there are no systematic approaches available.

3.3.6
Examples for the Use of Surfactants in Plant Extraction

3.3.6.1 Plant Extraction Using Micellar and Cloud Point Extraction

The micellar extraction is used for the extraction of phyto compounds from plant [50–53] or for liquid–liquid partition in further downstream processing. Paclitaxal from *Taxus chinensis* is removed from the plant with a MTBE–hexane mixture. The target substance in this organic crude extract is purified by a partition into a aqueous micellar phase [54]. Cloud point extraction is a well known technique for the separation of proteins [34, 38] and other biomolecules [37, 55]. Some examples in Table 3.4 show that cloud point extraction is used as a powerful tool for a preconcentration of the target substance in the surfactant-rich phase [50, 52, 53, 56–58]. Müller et al. [59] selected commercially available surfactants with a high potential to form mesophases and the ability to extract natural compounds from plants. Bart et al. [60] show the feasibilty of a process for the extraction of triterpenic acids with mesophases. Müller and Triantafillaki [61] focused on the extraction of natural indigo colorants and estimated a partition coefficient of minimum 15 for the red colorant indirubin as well as a mass transfer coefficient for the transfer from the aqueous to the mesophase of the Tween–water systems used. The back extraction can be avoided if the whole mesophase loaded with the active agent can be used as an ingredient for secondary products, (e.g., cosmetic creams) [61].

As already mentioned Bordier [38] reported the first application of cloud point extraction and determined the partition of hydrophobic and hydrophilic proteins in a two-phase system of Triton-X114 solution in water (2%w/v) above cloud point temperature. The results show that hydrophobic proteins are recovered with the small surfactant-rich phase while the hydrophilic ones are mainly found in the water phase. 85% of cytochrome b5 is recovered with the surfactant phase. For catalase a recovery of 90% with the aqueous phase is reported. The extraction in aqueous micellar two-phase systems is controlled by hydrophobic interactions between the target proteins and the surfactant. For cholesterol oxidase the solubility in the surfactant-rich phase increases by changing the ionic strength (addition of salt) or increasing the temperature [34]. Polyphenol oxidase from grape berries [56], mushrooms [57] and broad beans [62] can be prepurified by cloud point extraction. The enzyme is recovered with the aqueous phase while chlorophylls and polyphenols [56] are removed by the surfactant phase. Figure 3.19 exhibits the extraction process for latent mushroom tyrosinase. The cleaned mushrooms are treated with a buffered aqueous solution. After separation of the solids the surfactant is added twice to the solid-free solution to separate the hydrophobic proteins and phenols. Finally, the target enzyme is precipitated by addition of ammonium sulfate.

Table 3.4 Application of micellar and cloud point extraction in plant extraction.

Extracted Molecule	Surfactant System	Carrier matrix	Application	Reference
Tyrosinase (polyphenoloxidase)	Triton X-114	Aqueous extracts from mushroom (*Agaricus bisporus*) Grape berries	Purification by cloud point extraction	[54, 55]
Anthraquinone derivatives	Genapol-X080/water	Rhubarb *Morinda citrifolia*	Micellar extraction from plant and preconcentration by cloud point extraction at 55 °C and 0.24 g mL^{-1} sodium sulfate	[48, 51]
Cryptotanshinone, tanshinone I, tanshinone IIa	Genapol-X080/water	*Salvia militiorrhiza bunge*	Micellar extraction from plant	[49]
Paclitaxal	*N*-cetyl-pyridinium-chloride–water	MTBE/hexane based extract from *Taxus chinensis*	Prepurification by l-l extraction from organic solvent	[52]
Ginsenosides	Triton X-100	Ginseng root	Pressurized micellar extraction from the root and preconcentration by cloud point extraction	[50]
Saponins (glycyrrhizic acid), Flavonoids (liquiritin)	Triton X-100	Licorice root	Micellar extraction from root and cloud Point extraction	[56]
Polyphenols Tocopherol	Genapol-X080/water	Olive mill wastewater	Recovery and concentration by cloud point extraction at 55 °C	[53]

Our own results show the high potential of the cloud point extraction for the purification of hydrophilic oil seed proteins with Genapol-X80. Colored substances and polyphenols are removed by the surfactant-rich phase and the protein solution is purified. Gortzi *et al.* [55] use cloud point extraction with Genapol-X80 for the removal of polyphenols and tocopherol from olive oil mill waste water. With 5% surfactant a polyphenol recovery of ca. 80% and α-tocopherol recovery of more

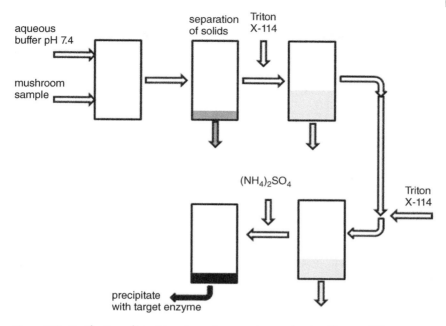

Figure 3.19 Purification of latent tyrosinase from mushroom *Agaricus bisporus*. [57]

than 99% is possible. Much higher concentrations of Genapol X-80 are not desirable because this leads to a strong increase in volume of the surfactant phase. These examples show that cloud point extraction can be used for a further refining of a product by extraction of impurities into the surfactant-rich phase which is discarded. Micellar extraction followed by cloud point extraction is usable for a recovery and preconcentration of plant compounds like anthraquinones [50, 53], ginsenosides [52] and saponins or flavonoids [58]. The addition of salt is used to affect the volume ratio of the two phases (surfactant rich and surfactant poor). With increased salt concentration the volume of the surfactant-rich phase is reduced and thus the preconcentration factor increased. Furthermore the target solutes become less soluble in the surfactant poor phase [50]. Thus in cloud point extraction high recoveries can be reached. For glycyrrhizic acid and liquiritin a NaCl concentration of $0.3\,g\,ml^{-1}$ leads to a preconcentration of 13.5 and recoveries above 98% for both substances [58]. For anthraquinone derivatives with addition of $0.25\,g\,ml^{-1}$ Na_2SO_4 a preconcentration of 12.5 with recoveries for the different substances from 75 to nearly 100% is reported [50]. Additionally this example shows that micellar solutions can compete with traditional organic solvents. A solution with 10% Genapol is more efficient than many polar or nonpolar solvents and as efficient as methanol [50].

The micellar and cloud point extraction of the triterpenic acids oleanolic and ursolic acid with Genapol X-80 is strongly affected by the pH (see Figure 3.20). At

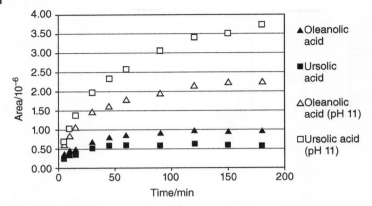

Figure 3.20 Extraction of ursolic acid and oleanolic acid from *Salvia triloba*. [61]

constant pH 11 the surfactant-rich phase contains more ursolic than oleanolic acid. The extraction with variable pH leads to extracts enriched in oleanolic acid. Thus, selectivity of the micellar extraction is strongly influenced by variation of pH [61]. The comparison of the HPLC peak areas (corresponds to concentrations) shows that a high pH leads also to an increase in the total amount of extracted triterpenic acids.

3.3.6.2 Plant Extraction Using Reverse Micelles

The examples (see Table 3.5) for the extraction of biomolecules presented in this chapter show that reverse micellar extraction is applied to protein and enzyme extraction or prepurification [63–69]. Reverse micellar solutions can be used in the first solid–liquid extraction (leaching) step [68, 69] and are more often applied for prepurification of an aqueous crude extract produced by a conventional solid–liquid extraction with water [63–66].

Reverse micelles (RM) are applicable in the simultaneous removal of triglycerides, proteins and glycosinolates [69] and other minor compounds like chlorogenic acid [70] from oil seed meal. The RM solutions are prepared by solubilization of a surfactant in an alcohol like isopropyl alcohol or *n*-butanol and addition of a nonpolar solvent to this blend. As nonpolar solvents hexane, isooctane or cyclohexane can be used. Finally water is added to obtain the desired ratio of water to surfactant (W_0). The removal of oil and protein is carried out by an immersion extraction using a suspension of seed meal in the RM solution. The duration of this step is about 1 h. The back extraction is performed by adding water, or an AOT–water mixture, to the RM extract. After 30 min the organic and aqueous phase are separated. The organic phase contains the oil. Proteins and glycosinolates are found in the aqueous phase. The oil yield is nearly not affected by W_0 and is above 95%. The protein recovery increases with the water-to-surfactant ratio. Since the micelle size increases with higher water-to-surfactant ratio larger proteins are extracted by the RM solution. For the enzyme myrosinase with a

Table 3.5 Application of the reverse micelles in plant extraction.

Extracted Molecule	Surfactant System	Carrier matrix	Application	Reference
Bromelain	AOT/iso-octane CTAB/iso-octane/ hexanol/butanol	Aqueous pineapple extract	Purification by reverse micellar extraction	[61, 62]
Horseradish peroxidase	AOT	Dialysate from aqueous extract	Purification by reverse micellar extraction	[63]
β-galactosidase	AOT/iso-octane	Aqueous extract of barley	Primary purification by reverse micellar extraction	[64]
Polyphenol oxidase	DTAB/iso-octane	Aqueous crude extract	Partial purification by reverse micellar extraction	[65]
Soy protein Isoflavones	AOT/iso-octane	Soy meal	Enrichment of proteins with isoflavones by back extraction	[66]
Triglycerides, proteins (myrosinase), glycosinolates, chlorogenic acid	CTAB, AOT, Tween 85, Triton X-100 with n-butanol or isopropyl alcohol and organic solvent (hexane, iso-octane or cyclohexane)	Oil seed meal	Simultaneous extraction of oil and protein, extraction of minor components in oil seed	[67, 68]

Abbreviations: AOT: sodium bis 2(ethyl-1-hexyl) sulfosuccinate, CTAB: cetyltrimethylammonium bromide, DTAB: dodecyl trimethyl ammonium bromide.

molecular weight of ca. 140 kDa it can be shown that a W_0 of 40 is needed to reach high recoveries. Thus, a size-dependent fractionation of proteins is feasible (see Figure 3.21).

After deactivation of myrosinase removal of glycosinolate by percolation with a CTAB–isooctane–10%butanol and a W_0 of 10 is possible. For back extraction water is added and the glycosinolates are transferred to the aqueous phase. For this procedure a glycosinolate recovery of more than 80% is reported. Other surfactants like Triton X-100, Tween 85 and AOT show a much lower efficiency.

Another application of RM extraction in oil seed protein recovery is the production of soy protein isolates enriched with isoflavones [68]. The proteins and isoflavones are removed from the meal by an AOT–water–iso–octane system. The back extraction is performed with an aqueous buffered stripping phase. By variation

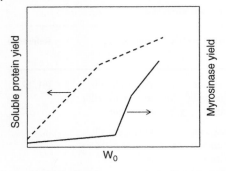

Figure 3.21 Recovery of water soluble protein and myrosinase dependent on water to surfactant ratio (RWS) as a qualitative plot. [12]

Table 3.6 Purifications and recoveries of enzymes at optimized conditions.

Target enzyme	Activity recovery	Purification	Total protein recovery	Reference
Bromelain	97.56%	4.54-fold	ca. 56% forward and 35% back extraction	[61, 62]
Horseradish peroxidase	80%	20-fold	–	[63]
β-galactosidase	98.74%	7.2-fold	13–14.4%	[64]
Polyphenol oxidase	78%	12-fold	<10%	[65]

of ionic strength, temperature and contact time it is possible to modify the isoflavone content of the proteins. Proteins extracted by reverse micellar extraction show differences in surface hydrophobicity, thermal rheological and gel properties compared with those produced derived from extraction with aqueous buffer solutions [71].

Further applications of RM solutions in plant extraction shown in Table 3.5 are secondary purification steps after a first aqueous (leaching) extraction. Polyphenol oxidase from apple skin [67] can be removed from an aqueous crude extract by a DTAB–iso-octane–hexanol system and is easily back extracted into a buffered aqueous ethanol solution. Under optimum conditions the target substance is 12-fold enriched and the maximal activity recovery is 78%. A high activity recovery can be reached by addition of AOT. The binding of counter-ionic surfactants like DTAB and AOT leads to a neutralization of the surfactant shell and a collapse of the reverse micelles. Other examples are the purification of β-glactosidase from barley [66], horseradish peroxidase from *Armoracia rusticana* root [65] and bromelain from aqueous pineapple or pineapple waste extracts [63, 64]. For high activity recoveries and purifications in the total process with forward and back extraction it is necessary to find optimum conditions. They depend on:

- type and concentration of surfactant and solvent,
- pH and ionic strength of the initial aqueous phase,
- solvent to co-solvent ratio for forward extraction,
- ionic strength, buffer pH, salt type and concentration for back extraction [63–69].

Furthermore, sometimes co-solvents like alcohols are needed to increase the micelle size [63]. A pH close to the isoelectric point can lead to interface precipitation due to electrostatic interaction of the target enzymes with the surfactant [66]. Table 3.6 gives an overview of results with the application of RM extraction. The comparison of activity and total protein recovery shows that RM extraction technique under optimum conditions may lead to a good selectivity for the target enzyme which results in purification factors up to 20.

3.4
Summary

Ionic liquids and surfactant systems are used in liquid–liquid as well as solid–liquid extraction processes. In the production of natural compounds the solid–liquid extraction is of importance for the removal of the target substance from the plant material. In the further downstream processing liquid–liquid extraction may play an important role. Ionic liquids can be used for the extraction of different substance classes from the biomass. This review shows their application in the recovery of alkaloids, sesquiterpenes, polyphenols, amino acids and proteins. In particular the artemisinin and lignin extraction exhibit the feasibility of new extraction processes based on ionic liquids. It is possible to design tailor-made solvents by the appropriate choice of anion and cation and by further chemical modification of the cation. This gives the opportunity to obtain high selectivities already in the first solid–liquid extraction step. Ionic liquids are often referred to as "green solvents" due to their low or negligible volatility. However, in a closed solvent extraction cycle the low volatility may lead to a disadvantage. Simple evaporation cannot be applied for the solvent recovery. Other methods have to be developed to make ionic liquids attractive as an alternative solvent on production scale.

Surfactants in aqueous solution are successfully used for the micellar extraction of biomolecules like proteins, polyphenols and triterpenes from plant and fungi. By an increase in temperature or addition of salts the cloud point is reached leading to a two-phase system. Dependent on its nature the target substance is either in the surfactant-rich phase or in the water phase. Because the surfactant-rich mesophase has a volume of 10 to 50% of the total volume, a concentration of the solute by the factor 2 to 10 is feasible. If hydrophilic compounds are processed the cloud point extraction is a tool for the purification of the solute because the more hydrophobic impurities can be removed with the surfactant-rich phase. Similar to ionic liquids surfactant molecules have a very low volatility which makes the solvent recovery and product removal from the extract more complicated than for typically used organic solvents. Micellar and cloud point extraction is well established in biotechnology for the recovery of membrane proteins. But for the extraction of natural compounds from plant economically feasible methods still have to be developed.

A further application of surfactants in the plant extraction is the use of reverse micelles formed in a system which consists of an organic solvent, a surfactant, and water. The surfactant builds a membrane around small water droplets dispersed in an organic solvent. These systems are not only used in the downstream processing of crude extracts but also in solid–liquid extraction. Particularly with regard to the purification of enzymes the reversed micellar extraction is of high relevance. On the other hand the simultaneous extraction of oil and proteins from oil plants points out the advantage of the reverse micelles in the removal of the target compounds from the biomass. The hydrophobic oil is recovered with the organic phase while the hydrophilic components are extracted by the water phase enclosed in the reverse micelles.

The applications presented in this contribution point out that ionic liquids and surfactant systems can successfully be used in solid–liquid extraction from biomass and the further downstream processing. If this new class of solvents exhibits a higher selectivity than traditional ones, processes based on the use of ionic liquids and surfactants may drastically improve the production processes of natural compounds.

References

1 Jin, I.J., Ko, Y.I., Kim, Y.M. and Han, S.K. (1997) *Arch. Pharm. Res.*, **20** (3), 269.
2 Braun, O. (2007) Zur Beziehung zwischen Struktur und Lösungsmitteleigenschaften von ionischen Flüssigkeiten mittels Headspace-Gaschromatographie. Dissertation, Friedrich Schiller Universität Jena.
3 Eßer, J. (2006) Tiefentschwefelung von Mineralölfraktionen durch Extraktion mit ionischen Flüssigkeiten. Dissertation, Universität Bayreuth, Shaker-Verlag Aachen.
4 Eßer, J., Jess, A. and Wasserscheid, P. (2003) *Chem. Ing. Techn.*, **75** (8), 1149.
5 Visser, A.E., Swatlowski, R.P. and Rogers, R.D. (2000) *Green Chem.*, **2** (1), 1.
6 Zhao, H. (2005) *J. Chem. Technol. Biotechnol.*, **80** (10), 1089.
7 Arce, A., et al. (2009) *Green Chem.*, **11** (3), 365.
8 Lapkin, A.A., Plucinski, P.K. and Cutler, M. (2006) *J. Nat. Prod.*, **69** (11), 1653.
9 Walker, A. (2007) WO 2007110637 A1, Oct. 4.
10 Lu, Y., et al. (2008) *J. Chromatogr. A*, **1208** (1–2), 42.
11 Du, F.Y., Xiao, X.H. and Li, G.K. (2007) *J. Chromatogr. A*, **1140** (1–2), 56.
12 Han, X. and Armstrong, D.W. (2007) *Acc. Chem. Res.*, **40** (11), 1079.
13 Li, M., Pham, P.J., Pittman, C.U. and Li, T. (2008) *Anal. Sci.*, **24** (10), 1245.
14 Lee, S.H., Doherty, T.V., Linhardt, R.J. and Dordick, J.S. (2009) *Biotechnol. Bioeng.*, **102** (5), 1368.

References

15 Tan, S.S.Y., et al. (2009) Green Chem., **11** (3), 339.
16 Cao, X., et al. (2009) Anal. Chim. Acta, **640** (1–2), 47.
17 Du, F.Y., Yiao, Y.H., Luo, X.J. and Li, G.K. (2009) Talanta, **78** (3), 1177.
18 Wang, J., Pei, Y., Zhao, Y. and Hu, Z. (2005) Green Chem., **7** (4), 196.
19 Wasserscheid, P. and Welton, T. (eds) (2008) Ionic Liquids in Synthesis, vol. 1, Wiley-VCH Verlag GmbH, Weinheim, Germany.
20 Seddon, K.R., Stark, A. and Torres, M.J. (2000) Pure Appl. Chem., **72** (12), 2275.
21 Widegren, J.A., Laesecke, A. and Magee, J.W. (2005) Chem. Commun., (12), 1610.
22 Favre, F., Olivier-Boubigou, H., Commereuc, D. and Saussine, L. (2001) Chem. Commun., (15), 1360.
23 Swatlowski, R.P., et al. (2002) Green Chem., **4** (2), 81.
24 Jastorff, B., et al. (2005) Green Chem., **7** (5), 362.
25 Gathergood, N., Scammells, P.J. and Garcia, M.T. (2006) Green Chem., **8** (2), 156.
26 Harjari, J.R., Singer, R.D., Garcia, M.T. and Scammells, P.J. (2008) Green Chem., **10** (4), 436.
27 Earle, M.J., et al. (2006) Nature, **439** (7078), 831.
28 Kuhn, T. and Wang, Y. (2008) Artemesinin–An Innovative Cornerstone for Anti-Malaria Therapy, in Natural Compounds As Drugs, vol. II (eds F. Petersen and R. Amstütz), Birkhäuser, Basel-Boston-Berlin, pp. 383–422.
29 Texter, J. (1999) Characterization of surfactants, in Surfactants–A Practical Handbook (ed. K.R. Lange), Hanser Publisher, Munich, Germany, pp. 1–3.
30 Shinoda, K., Nakagawa, T., Tamamushi, B.-I. and Isemura, T. (1963) Colloidal Surfactants, Some Physiochemical Properties, Academic Press, London, UK.
31 Holmberg, K., Jönssen, B., Kronberg, B. and Lindman, B. (2007) Surfactants and Polymers in Aqueous Solution, 2nd edn, John Wiley & Sons, Ltd, Chichester, UK.
32 Larsson, K., Lipids-Molecular Organization (1994) Physical Functions and Technical Applications, The Oily Press, Dundee, UK.
33 Müller, U. (1991) Extraktiv-Fermentation von Sekundärmetaboliten am Beispiel von Cycloheximid. Dissertation, Universität Dortmund/D.
34 Tani, H., Kamidate, T. and Watanabe, H. (1998) Anal. Sci., **14** (5), 875.
35 Li, J.L. and Chen, B.H. (2003) J. Coll. Interf. Sci., **263** (2), 625.
36 Werck-Reichhardt, D., et al. (1991) Anal. Biochem., **197** (1), 125.
37 Quina, F.H. and Hinze, W.L. (1999) Ind. Eng. Chem. Res., **38** (11), 4150.
38 Bordier, C. (1981) J. Biol. Chem., **256** (4), 1604.
39 Müller, U., Merrettig-Bruns, U. and Hollmann, D. (1995) Chem. Ing. Tech., **67** (6), 783.
40 Onken, U., Müller, U., Träger, M. and Heusch, R. (1988) DPA 3817957, 27.5.
41 Schürholz, T., Gieselmann, A. and Neumann, E. (1989) Biochim. Biophys. Acta, **986** (2), 225.
42 Mazzola, P.G., et al. (2008) J. Chem. Technol. Biotechnol., **83** (2), 143.
43 Hilhorst, R., et al. (1992) Pure Appl. Chem., **64** (11), 1765.
44 Wang, W., Weber, M.E. and Vera, J.H. (1995) Ind. Eng. Chem. Res., **34** (2), 599.
45 Nishiki, T., Sato, I. and Kataoka, T. (1993) Biotechnol. Bioeng., **42** (5), 596.
46 Lee, S.S., et al. (2005) Korean J. Chem. Eng., **22** (4), 611.
47 Buggert, M., et al. (2006) Chem. Eng. Technol, **29** (5), 567.
48 Mokrushina, L., et al. (2007) Ind. Eng. Chem. Res., **46** (20), 6501.
49 Niklas, Y.J., et al. (1992) Macromolecules, **25** (18), 4797.
50 Shi, Z., Zhu, X., Cheng, Q. and Zhang, H. (2007) J. Liq. Chromatogr. Relat. Technol., **30** (2), 255.
51 Shi, Z., Wang, Y. and Zhang, H. (2009) J. Liq. Chromatogr. Relat. Technol., **32** (5), 698.
52 Choi, M. (2009) J. Chromatogr. A, **983** (1–2), 153.
53 Kiathevest, K., et al. (2009) Sep. Purif. Technol., **66** (1), 111.
54 Jeon, K.-Y. and Kim, J.-H. (2007) Biotechnol. Bioprocess Eng., **12** (4), 354.

55 Gortzi, O., et al. (2008) *J. Am. Oil Chem. Soc.*, **85** (2), 133.
56 Sanchez-Ferrer, A., Bru, R. and Garcia-Carmona, F. (1989) *Plant Physiol.*, **91** (4), 1481.
57 Nunez-Delicado, E., Bru, R., Sanchez-Ferrer, A. and Garcia-Carmona, F. (1996) *J. Chromatogr. B*, **680** (1–2), 105.
58 Sun, C., Xie, Y. and Liu, H. (2007) *Chin. J. Chem. Eng.*, **15** (4), 474.
59 Müller, U., Triantafillaki, C. and Jordan, V. (2007) *Chem. Ing. Tech.*, **79** (9), 1363.
60 Bart, H.-J., et al. (2008) *Chem. Ing. Tech.*, **80** (9), 1292.
61 Müller, U., et al. (2009) FH^3-Project FKZ 1773A05. final report, TIB Hannover.
62 Bruins, R.H. and Epand, R.M. (1995) *Arch. Biochem. Biophys.*, **324** (2), 216.
63 Hemavathi, A.B., Hebbar, H.U. and Raghavarao, K.S.M.S. (2007) *J. Chem. Technol. Biotechnol.*, **82** (11), 985.
64 Hebbar, H.U., Sumana, B. and Raghavarao, K.S.M.S. (2008) *Bioresour. Technol.*, **99** (11), 4896.
65 Huang, S.-Y. and Lee, Y.-C. (1994) *Bioseparation*, **4** (1), 1.
66 Hemavathi, A.B., Hebbar, H.U. and Raghavarao, K.S.M.S. (2008) *Appl. Biochem. Biotechnol*, **151** (2–3), 522.
67 Imm, J.Y. and Kim, S.C. (2009) *Food Chem.*, **113** (1), 302.
68 Zhao, X., et al. (2008) *J. Sci. Food Agric.*, **88** (4), 590.
69 Ugolina, L., De Nicola, G. and Palmieri, S. (2008) *J. Agric. Food Chem.*, **56** (5), 1595.
70 Leser, M.E., Luisi, P.L. and Palmieri, S. (1989) *Biotechnol. Bioeng.*, **34** (9), 1140.
71 Zhao, X., et al. (2006) *Food Chem.*, **111** (3), 599.

4
High Pressure Processing

Rudolf Eggers and Stephan Pilz

4.1
Introduction

For more than 30 years extraction using supercritical carbon dioxide ($scCO_2$) has been the most prominent application of supercritical fluids [1]. Originally, the focus was on the refinement or recovery of raw materials in the food industry, for example, decaffeination of coffee or production of hop and spice extracts [2]. However, recently, further applications for supercritical fluids have been developed [3]. For various reasons supercritical fluids had their first application in the food industry. In respect to the design of an extraction process several aspects are essential:

- The solubility of the substance to be separated in the solvent;
- The selectivity compared with other substances;
- The diffusion in the case of solid raw materials (capillary effects);
- The effort in removing the solvent after extraction;
- Regulatory approval of the solvent within the application range of the retrieved products.

Supercritical fluids combine the physical properties of fluids and gases in an advantageous way, so that they exhibit high solvent power and high density, together with fast diffusion, low surface tension and low viscosity. The combination of high solubility and low capillary action is especially attractive to extraction from solid raw materials when the target substances are deep inside pores. Since physical properties can be manipulated over a wide range by pressure and temperature, the selectivity can be controlled.

In $scCO_2$ those substances are well-soluble which posses low molecule weight, such as alcohols, aldehydes, ketones, esters and some alkaloids. These substances are the basis for flavors. Less well-soluble in $scCO_2$ are substances with high molecular weight, such as amino acids, glycosides, proteins and carbon hydrates. These substances are related to functionality. The solubility can be manipulated by additives, usually ethanol and water. Furthermore, processes using a mixture of gases are possible. For example, a mixture of carbon dioxide and propane combines low capillary resistance with high solubility, respectively.

Constantly increasing demands on product quality, especially on purity, demanded by end consumers, as well as increasing attention to safe and environmentally friendly production forced the World Health Organization (WHO) and also the European Union to publish guidelines for recommended solvents in food production. The list of substances is being adopted by an increasing number of governments. The following substances are classified as harmless solvents [4]: propane, butane, ethyl acetate, butyl acetate, ethanol, CO_2, acetone, N_2O.

All these substance have to be provided at "food grade" to be applicable. So for example, ethanol must not be denatured, propane must exhibit a purity of at least 95%, and carbon dioxide must not have been produced technically, but instead be gained from natural mineral water springs. For cosmetics, health, and pharmaceuticals similar requirements apply.

At first glance, extraction using $scCO_2$ does not seem to be competitive with conventional extraction with ethanol, because high pressure seems to lead to higher investment costs and to higher operational costs related to energy. Additionally, it seems to cover more process steps. A more detailed study might lead to different results. So, in both cases for solid material extraction, autoclaves have to be filled, the solvent has to pass through a packed bed and afterwards the solids have to be removed. In the case of the high pressure extraction the process steps of pressurization and depressurization have to be added. Additionally, the investment costs of the autoclaves and of the peripheral equipment are higher compared with conventional extraction. On the other hand, the reconditioning, that is, the separation of the solvent from extracted substances, is far simpler for the supercritical process than for the conventional route. By depressurization the solvent power of the supercritical fluid drops and the retrieved extract is present in ultrapure form. In comparison, for conventional processes often a complex and costly post-treatment by thermal unit operations follows, since the gentle treatment of the product requires either multistage or processing vacuum. If the advantages of the simpler post-treatment compensate the higher investment costs, this is not usually sufficient to direct the decision to replace a conventional process by one using $scCO_2$. The decision is more often likely to be influenced when new product qualities can be gained, not solely reached by higher purity regarding the solvent but also by higher selectivity. Furthermore, the yield often turns out much higher due to the high accessibility of the pores. This counterbalances the more complex extraction cycle. One further advantage of using $scCO_2$ is the low process temperature of about 40 °C, at which high pressure extraction can be performed. Due to the low thermal load, product changes, especially for food applications, can be excluded. It is not possible to make a general recommendation as to which product systems benefit from a high pressure extraction; instead this has to be proven case by case. Certainly, also production aspects regarding environment and safety play a major role, as well as multiproduct plant considerations.

Based on worldwide applications of supercritical fluids in production for a wide range of products the use of supercritical gases can be considered as state-of-the-art. Up to the end of 2004 roughly 90 production plants have been installed which use $scCO_2$ as solvent. The three most important contractors are in descending

Table 4.1 Distribution of industrial scale SFE plants over world regions by end of 2004.

Region	Number of installed SFE plants
Europe	45
Americas	5
Japan	17
China	9
Korea	3
India	3
Australia, New Zealand	3
Africa	1

order according to the number of constructed plants: Uhde High Pressure Technologies (Hagen/D), Natex (Ternitz/A), and Separex (Champigneulles/F). Some plants were engineered by the owners themselves and only components were purchased. Table 4.1 gives a distribution of the existing supercritical fluid extraction (SFE) production plants across the world [5]. For new plants two trends can be identified: pressure level is raised to 500...750 bar – compared with standard numbers of 300 . . . 350 bar – to allow higher yields, and the number of products as well as the diversification increases constantly.

Extraction with $scCO_2$ has been established for a series of applications. Two different targets can be distinguished:

1) Extractions for removal of a substance to refine the raffinate, for example:
    - decaffeination of tea,
    - decaffeination of coffee,
    - defatting of cacao.

Usually the raffinate is by far larger in volume. In some cases the extract also has a value and can be marketed.

2) Extractions to gain the extract as desired product, for example:
    - recovery of hop extracts,
    - fruit flavors,
    - citrus flavors,
    - spice flavors,
    - vanilla, peanut, hazelnut, etc.,
    - coffee and cacao flavor,
    - extracts of milk products,
    - essential oils,
    - fish oil components,
    - flavor extracts from alcoholic beverages.

The high product concentration is often unique. Selectivity and high yield allow reasonable prices. The vitality of SFE use for industrial applications is also indicated

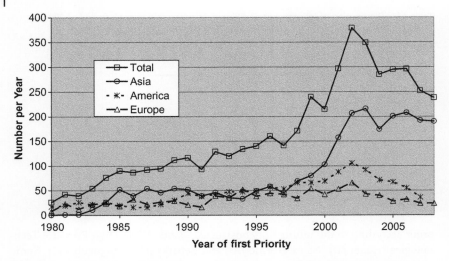

Figure 4.1 Patent applications concerning supercritical extraction per year for several world regions.

by the expanding number of patents. Figure 4.1 shows the number of patents filed over time for several regions of the world [6].

Besides SFE applications further processes using supercritical fluids have been developed within the last 20 years. They all benefit from the strong adjustability of the solvent power. Consequently, supercritical fluids are exploited in particular areas [7] (those are reviewed elsewhere):

- SFE on an industrial and analytical scale [8, 9].
- Chromatography and fractionation [10, 11].
- Reactions [12, 13].
- Formation of microparticles [14, 15].
- Purification of high value products [8].
- Spray painting and dyeing [16].
- Polymer synthesis and processing [17].

In most cases, CO_2 is the supercritical medium; however other substances are also used, for example, propane, ethanol, ammonia. The most extreme behavior is exhibited by water when it reaches the supercritical state.

4.2
Supercritical Fluids

The underlying interest in supercritical fluids has arisen because they can have properties intermediate between those of typical gases and liquids. The conditions may be optimum for a particular process or experiment. Furthermore, properties are controllable through both pressure and temperature, and the extra degree of freedom can allow more than one property to be optimized. However, any advantage has to be weighed against the cost and inconvenience of the higher pressure needed.

4.2.1
General

The standard phase diagram in Figure 4.2 depicts the regions of temperature and pressure wherein the substance is present as solid (S), as liquid (L), vapor (V) and as gas (G). The phases are separated by the individual phase-transition lines, at which the phases are in equilibrium [18]. All three (standard) phase boundaries meet at the triple point (TP), where, the three phases coexist and are in equilibrium. Also, the generalized diagram shows the region of supercritical fluid (SCF) apart from the three standard aggregate states.

Moving upwards along the gas–liquid coexistence curve, which is a plot of vapor pressure versus temperature, both temperature and pressure increase. The liquid becomes less dense because of thermal expansion, and the gas becomes denser as pressure rises. At the critical point (CP), the densities of the two phases become identical and the distinction between the gas and the liquid disappears. (This was first discovered by Baron [19].) Beyond this point, the substance can only be described as a fluid: the P–V supercritical fluid.

Consulting a P–V diagram (s. Figure 4.3), this becomes even more evident: while heating up a saturated substance means moving up the boiling line in the P–T diagram, in the P–V diagram it describes the isotherms climbing and correspondingly increasing the pressure on both the saturated liquid line and the saturated vapor line. For elevating conditions, the differences in density between the two lines declines and finally disappears when the CP is approached. This means, from a molecular perspective, that the intermolecular forces become equal and, from an energy view, the thermal energy becomes equal.

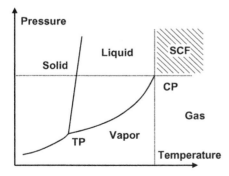

Figure 4.2 Generalized qualitative P–T phase diagram.

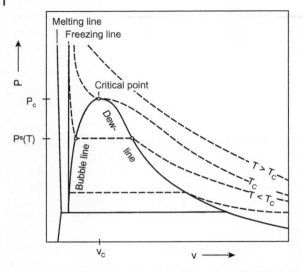

Figure 4.3 Qualitative P–V diagram with isothermals and phase-transition lines.

The isotherm of the critical temperature touches the dew point line and the bubble point line at the CP, exhibiting a saddle and a turning point. Similar curves and mathematical relations can be also derived from other two-phase diagrams. What they all have in common is that the differences between liquid and vapor phase are growing smaller and smaller. Hence, the CP is also defined by the disappearance of the heat of vaporization.

Another key feature of this diagram is the indication that, within the supercritical region, there is no phase boundary between the gas and the liquid phases, that is, there is a continuity in physical properties of the fluid between the gas and the liquid states. The consequence is that supercritical fluids have properties which are a curious hybrid of those normally associated with liquids and gases. Thus, under most conditions, the viscosities and diffusivities are similar to those of gases, while the density is closer to that of a liquid. The CP is characterized by its temperature and pressure coordinates, which are very individual for different substances. In Figure 4.4, they are plotted for several fluids [20] which have already been utilized in technical plants using supercritical fluids (H_2O, CO_2, C_2H_4). Also, chemically similar fluids are compared, for example, characterized by their bonding (C_2H_2, C_2H_4, C_2H_6) or by their molecule lengths (C_2H_6, C_3H_8, C_4H_{10}, C_5H_{12}).

The two most popular fluids are carbon dioxide and water. Since they are non-toxic and nonflammable, they are essentially environmentally benign solvents that can be used – even for food processing – without significant regulation. In addition, they are two of the most inexpensive solvents available commercially.

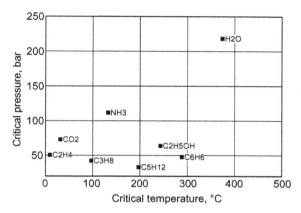

Figure 4.4 Critical points of some fluids.

Table 4.2 Order of magnitude of density, diffusion coefficient and dynamic viscosity for gases, liquids, and SCFs.

Physical property	Symbol	Unit	Gas	Liquid	SCF
Density	ρ	gcm^{-3}	10^{-3}	1	0.2...0.8
Diffusion coefficient	D_{12}	$cm^2 s^{-1}$	10^{-1}	10^{-6}	10^{-3}
Dynamic viscosity	η	$gcm^{-1} s^{-1}$	10^{-4}	10^{-2}	10^{-4}

4.2.2
Physical Properties

A unique feature of SCFs is their pressure-dependent density. If the temperature is constant, density can be adjusted by pressure from that of vapor to that of a liquid with no discontinuity. Several other properties change simultaneously with this change in density: those more relevant to reactions include viscosity, solvent power, diffusivity, and dielectric constant. The sensitivity of all these properties is highest in the vicinity of the CP [21]. It is common to focus on the region where the reduced temperature $T_r = T/T_c$ and reduced pressure $P_r = P/P_c$ are of the order of unity. In this region, considerable changes in fluid density and related properties such as solubility are observed with small changes in pressure. These characteristics make SCFs very attractive as tunable process solvents or reaction media. Table 4.2 demonstrates, for three properties, how an SCF fills the gap between the liquid and gas states and how it can behave as both liquid and gas [22].

Undoubtedly, the pivotal tunable property of an SCF is its solvent power. It features many chemical reactions and processes where reduction in pressure and associated expansion of the solvent is an efficient method for recovering solid products from supercritical reactions. Also of particular significance is the almost complete miscibility of permanent gases in SCFs. One of the most unexpected properties of SCFs is their ability to dissolve solids to a high extent compared with gases. The good solubility of SCFs enables processes with solid or gaseous educts in homogenous phase, for example, reactions. All these characteristics make SCFs very attractive as tunable process solvents or reaction media.

4.2.3
Solvent Power and Solubility

The solvent power of supercritical fluids is, as described above, rather more liquid-like than gas-like. However, besides the pressure effect it still depends on the molecule, its size and structure. So, polarity is a major issue and not all substances exhibit the same solubility in compressed gases. Figure 4.5 gives an overview by showing the solubility of selected substances in three compressed gases: CO_2, propane and DME. The last two are less often used in industrial applications. As a basic rule it can be stated that CO_2 preferably dissolves apolar substances but does not behave as apolar as propane since its quadruple momentum gives it some polarity.

Since in many cases the solvent power of the SCFs is not sufficient—even if the pressure is further elevated—co-solvents are added. Hereby, the polarity is manipulated and the solubility becomes influenced. The two most typical co-solvents are

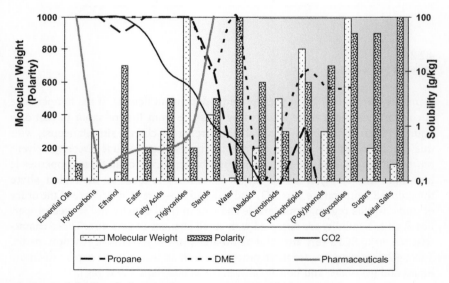

Figure 4.5 Solubility of selected substances in compressed gases.

ethanol and water. Both shift the polarity of the system and enable the extraction of polar components. Additionally, they belong to the list of harmless solvents. The concentration and selection of the co-solvent depends on the substances to be extracted. As a rule of thumb, the amount of added co-solvent should not exceed 10 wt.%, otherwise each of these solvents could be used on its own. This statement only considers the solubility during extraction step and does not cover other advantages of an SFE application like simple solvent removal. In some cases antisolvents are used, either to suppress solubility or within a separation step. The use of co- and/or antisolvents provide even more options for SFE applications comprising total extraction, selective extraction, two-step extraction, and fractionation.

4.3
Physical Properties – Mass Transfer Characteristics

The process of supercritical extraction of natural substances from plant material is normally carried out as a batch operation. Thus the supercritical solvent flows through a solid bed in a vessel. If the solid bed offers a sufficient void fraction the supercritical solvent permeates without high pressure loss because the viscosity of CO_2 is very low even at high pressures. Moreover the diffusive transport of the fluid into the natural matrix is high. However, the resistance at unbroken cell membranes must be overcome in order to dissolve the desired extract component from the natural structure. Thus two types of mass transfer characteristics exist in supercritical treatment (Figure 4.6):

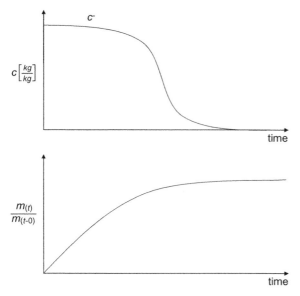

Figure 4.6 Mass transfer characteristics for supercritical extraction of solid natural material.

- A linear extraction profile of extracted mass with extraction time. The concentration of solute in the fluid remains constant following the phase equilibrium between the dissolved constituent and the supercritical fluid dependent on the process pressure and temperature. These extraction profiles are similar to those from leaching processes. The length of the linear profile depends on the efficiency of the pre-treatment process. The optimum is accomplished when the steady-state extraction remains until the total extractable mass has been transferred into the solvent.
- Normally the slope of extraction kinetics decreases before the natural material is extracted in total because the accessibility within the microstructure is incomplete for the solvent. Thus the second part of extraction kinetics is characterized by high diffusive resistances in the natural matrix and additionally by the diffusion of the dissolved solute back to the surface of the solid particle.

Natural material that can be disintegrated by proper pre-treatment in order to get valuable extract components elucidate at first a linear part, based on the valid phase equilibrium and a subsequent second decreasing extraction profile, that is diffusion controlled. The length of the linear part demonstrates the efficiency of the pre-treatment process. Well-known examples of these combined kinetic profiles are the extraction of natural oils from crushed oil seeds [23–25], extracts from hops and spices or even oleoresins from paprika [26]. However, if the aim of the supercritical process is the extraction of disturbing constituents out of the natural matrix without any disintegration, the inner mass transfer resistance becomes dominant. The resulting extraction profile only reveals the decreasing diffusive controlled character as in Figure 4.6. Examples that represent this category of natural substances demand long extraction times, like the decaffeination of green raw coffee beans [27] and tea leaves, or lignin from wood [25].

A comparatively large number of approaches for modeling the mass transfer in supercritical processing of natural material has been carried out [24]. Often these models are based on experimental results that are derived from small laboratory devices and their validity in order to specify an up-scaled process remain questionable. The complex matter of modeling the mass transfer in supercritical processing of natural material requires at least the knowledge of the following items:

- The equilibrium solubility data in a broad range of pressure and temperature; even the concentration of the extractable component may influence the equilibrium data, this will lead to extract composition changes with the degree of extraction [28].
- The method of pre-treatment determines the degree of access ability for the solvent to the desired substances within the natural matrix. The method of preparation which ruptures as many cell walls as possible yields the highest extraction rates. Relevant parameters are both the inner and outer porosity. Thus, mass transfer modeling needs the particle size distribution and the void fraction of the solid bed in the extraction vessel. The particle size is related to

the interfacial area and the void fraction is responsible for a complete mass transfer contact of the solvent flow.

- The calculation of mass transfer coefficients based on Sherwood equations (Sh = f [Re, Sc]) that are valid for the superficial solvent velocities. Many of these correlations are existing, for instance from food drying processes [29]. However, the Reynolds Number often changes along the process because the void fraction of the solid bed in the vessel does not remain constant.
- The availability of an effective diffusion coefficient for the solvent in the porous structure is important in order to enable a reliable calculation of the diffusive controlled mass transfer during the second part of the extraction process. Often this transport property is just fitted to experimental results thus the general evidence is not provided. Moreover, effective diffusion coefficients may change during the extraction due to increasing inner porosities of the natural material.
- Depending on the ratio of vessel length to inner diameter and even on the homogeneity of void fraction a back mixing of the solvent flow may occur that decreases the mass transfer kinetic. In consequence a mathematical model actually has to consider at least an axial dispersion coefficient.
- Normally, no assumption is made about the distribution of the extract components within the particles of natural material.

The concentration i of an extracted natural substance in the supercritical fluid phase is related to the height z in a vessel filled with a solid bed consisting of pre-treated natural material. The balance equation is based on the non steady state differential equation that combines the convectional and the diffusive transport as well. With the assumptions of one dimensional piston flow and constant fluid density the differential equation for the solute concentration c in the supercritical fluid reads [30, 31]:

$$\frac{\partial c}{\partial t} = D_{ax}\frac{\partial^2 c}{\partial z^2} - u\frac{\partial c}{\partial z} + \frac{k \cdot a_z}{\varepsilon}(c^* - c) \qquad (4.1)$$

where $D_{ax}\,[m^2/s]$: axial dispersion coefficient;
$k\,[m/s]$: mass transfer coefficient from solid particle surface into the bulk fluid;
$a_s\,[m^2/m^3]$: mass transfer area per unit volume of solid bed;
ε: void fraction of solid bed;
$u\,[m/s]$: interstitial velocity of supercritical fluid;
$c\,[kg/kg]$: extract concentration in supercritical fluid;
$c^*\,[kg/kg]$: extract concentration at particle surface (equilibrium assumed);
$t\,[h]$: extraction time.

The equation describes the complete inner surface of the solid bed as wetted by the supercritical fluid. Thus the void fraction – the outer porosity of the solid bed – gives the volume of fluid between the particles. In the case of non-regular structures, which is evident after disintegration of the inner structure, this parameter has to be determined by experimental methods, like mercury porosity or the adsorption of gases. Moreover, the above differential equation states a constant remaining porosity during the extraction process, which does not actually exist for the reason of mass transfer by pore diffusion.

For simplification the particles are often considered to be spherical. In the case of a solid bed consisting of particles with uniform diameter d_p the overall mass transfer area a_s results in:

$$a_s = \frac{6 \cdot (1-\varepsilon)}{d_p} \tag{4.2}$$

Modeling of extraction based on Equation 4.1 neglects both the change of the bed porosity and the particle diameter. However, the important transport parameters in the differential balance equations are the mass transfer coefficient k and the axial dispersion coefficient D_{ax}. The mass transfer coefficient is a measure of the resistance to mass transfer from the solid surface to the fluid. The parameters responsible for influencing the transfer resistance are the thicknesses of the boundary layers for velocity and diffusion. Thus the mass transfer coefficient is a function of the fluid velocity u, density ρ, dynamic viscosity η and then diffusion coefficient D of the solute in the fluid. Based on the boundary layer theory numerous equations are available which predict the mass transfer coefficient using dimensionless numbers [29]. These correlations normally consider external flow round spherical shape of particles. Because these equations are developed at near ambient conditions their application in supercritical processing of natural material has yet to be proven.

As the velocities within the porous structure of the solid bed are relative low, the mass transfer by forced convection, expressed by the Reynolds number Re and the Schmidt number Sc, and may be superposed by natural convection expressed by the Grashof number (Sh = f[Re,Sc,Gr,ε]).

The axial dispersion coefficient in Equation 4.1 is a measure of a possible mixing effect while the fluid is moving through the porous solid bed. Boundary limit cases are plug flow without mixing ($D_{ax} = 0$) and infinite for total mixing ($D_{ax} = \infty$). It is evident that the axial dispersion coefficient mainly depends on the fluid flow velocities, the particle diameter and the homogeneity of the porosity. Although there are correlations given in literature for packed beds [18], the influence of mass transfer on dispersion is usually neglected. Severe effects, like fluid channeling in an inhomogeneous bed, will generate dramatic losses in bed efficiency and is not reflected by the dispersion coefficient.

Following the extraction profile given in Figure 4.6 the diffusive transfer of the solute in the porous structure of the particles is dominant at least in the final part of an extraction process. Thus the inner resistance has to follow an effective diffusion coefficient D_{eff} and the inner porosity ε of the particles. With a spherical particle the mass balance of the substance to be extracted corresponds to the pore fluid solute y_p and the particle solute phase concentration of the solute x_p

$$\varepsilon_p \frac{\partial y_p}{\partial t} = D_{eff} \cdot \frac{1}{r^2} \frac{\partial}{\partial r}\left(r^2 \frac{\partial y_p}{\partial r}\right) - \rho_p \cdot \frac{\partial x_p}{\partial t} \tag{4.3}$$

where $D_{eff}\,[m^2/s]$: effective diffusivity;
$y_p\,[kg/m^3]$: pore fluid concentration;
x_p: particle phase concentration expressed as mass ratio;
ε_p: particle porosity;
$r\,[m]$: particle radial coordinate;
$\rho_p\,[kg/m^3]$: particle density.

The most important parameters in Equation 4.3 are the effective diffusion coefficient D_{eff} and the mass transfer rate $\frac{\partial x_p}{\partial t}$ which is linked to the mass transfer coefficient k in Equation 4.1. As for the calculation of effective diffusion coefficients, simple equations have been proposed in literature [31, 32].

4.4 Process Units

4.4.1 Pre-Treatment

Extraction yield and product quality are highly dependent on pre-treatment steps that the natural material undergoes before the actual extraction process starts. Generally, the process units of conditioning solid material are classified as mechanical, thermal and biological methods of treatment. These process units are reasonable in order to simultaneously homogenize the feedstock and improve the mass transfer operations related to substances of the natural feedstock that are the desired product components of the extraction process. Mass transfer should be favored by maintaining not only small particles but also free access of the supercritical solvent to the desired natural substance within the natural matrix. As an example Figure 4.7 demonstrate that lipid cells in oil-bearing natural material are well protected [33].

Thus cell walls must be broken and further the outer structure of the solid bed should be porous and as homogeneous as possible. Large particles result in high void fraction, but at the same time the mass transfer inside the particles is delayed and obstructed. In terms of mass transfer the pre-treatment should promote efficient inner diffusive transport and outer convective transport. In consequence the relation of both transport effects during supercritical extraction, expressed as the Biot number for mass transport should be in the range of one.

$$Bi = \frac{\beta \cdot d_p}{D_{eff}} \approx 1 \tag{4.4}$$

where $d_p\,[m]$: particle diameter;
$\beta\,[m/s]$: mass transfer coefficient;
$D_{eff}\,[m^2/s]$: effective diffusion coefficient in natural material.

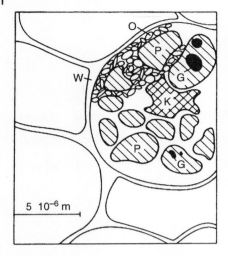

Figure 4.7 Structure of oil bearing natural material. G: globoid; K: nucleus; O: oil droplet; P: protein; W: cell wall [31].

Moreover, solid particles sometimes agglomerate during extraction by high moisture content although the material has been conditioned to small particles prior to extraction. The resulting decreased void fraction leads to channeling of the fluid within the solid bed or even blocking of the flow in the case of high pressure differences. As a consequence the extraction proceeds nonhomogeneously and its yield remains at low level. With respect to the product quality often disturbing components have to be considered by proper conditioning methods. For example certain enzymes must be deactivated by thermal pre-treatment. Innovative methods are part of current research: biological processes use the activity of proper enzymes in order to crack mass transfer resistant membrane material [34, 35]; high pressure gassing seems to be a promising method for swelling effects that widen pore diameters [36], and even electric field treatment using high frequent pulses are effective in disintegration of cell membrane walls in natural material [37]. Table 4.3 summarizes the different possibilities of material explaining the working principles, the technical methods and the achievable effects.

4.4.1.1 Mechanical Pre-treatment

Similar to conventional liquid extraction, the solid feed material has to be conditioned properly. Mass transfer should be favored by small and open diffusion paths throughout the broken particles. If required, skins and hulls of the natural material have to be removed by peeling or even separating in cyclone devices. Different equipment for mechanical pre-treatment exists due to the type of force applied. Cutting, crushing, milling and grinding or flaking is achieved by normal forces often combined with shear forces in profiled and smooth rolling mills. The

Table 4.3 Pre-treatment of natural material.

Pre-Treatment	Principle	Method	Effect
Mechanical	Disintegration	Grinding	Short diffusion distances
		Milling	Reduction of mass transfer Resistance
		Crushing	
		Breaking	
		Disruption	
	Shape forming	Decortication	Cell breaking
		Rolling	Uniform geometry
		Flaking	Homogeneous structure
		Screwpressing	
		Extruding	Low pressure loss during extraction; Uniform solvent flow
		Pelletising	
Thermal and hydrothermal	Adjustment of humidity	Drying	Open pores for extraction Inactivation of enzymes
		Wetting	Reduction of friction
		Swelling	Increase pore diameters
	Temperature increase	Steam treatment	Mobilizing of desired components Decrease in viscosity
			Expansion and destruction of pore
High pressure gassing	CO_2 impregnation	Expansion	Cell disrupture; micronizing
		Swelling	Increase of free volume
Biological	Enzyme activity	Fermentation	Reduction of undesired components
Electric fields	High frequency electrical pulses	Membrane disruption	Reduction of mass transfer resistance

working principle of a fluted breaker rolling mill that generates particles by ribbing is given in Figure 4.8.

Often roller mills work by compressing the material through small adjustable gaps between two rotating cylinders. If there is no differential circumferential speed, the natural material is pressed into thin flakes. Slight differential speed superposes shear forces and the material may also be ruptured. Moreover friction forces are applied in extruder devices and screw presses [38]. As an example Figure 4.9 shows the highly effective material conditioning of rapeseed with respect to

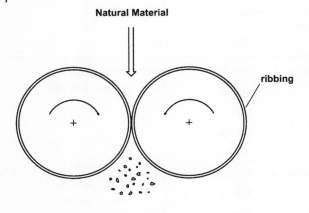

Figure 4.8 Working principle of a fluted breaker rolling mill.

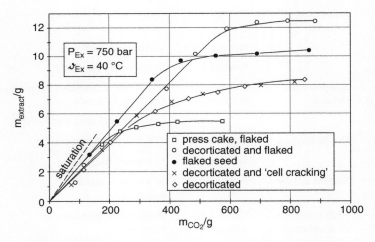

Figure 4.9 Flaking and screw pressing prior to supercritical extraction.

both methods, supercritical extraction of flaked material or press cake of screw pressed rape seed as well [33].

The yield of de-oiling nearly follows the dashed line representing the saturation loading which is based on the phase equilibrium between the extract and the supercritical solvent. However, the mass transfer resistance of hulls is high. Figure 4.10 shows the improvement in extraction yield by decorticating of seed hulls [38].

Often disintegration, for instance milling or grinding, leads to high pressure losses for the flow of solvent although the bulk densities are not very high. Moreover

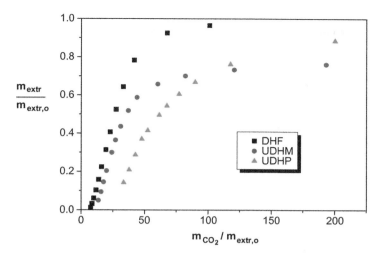

Figure 4.10 Decorticating prior to supercritical extraction.

nonhomogeneous extraction may increase channeling and at least the process reveals ineffective considering aspects of economic and quality. These disadvantages may be overcome by forming the disintegrated natural material into small cylindrical pellets by pressing the material through discs which are equipped with bore holes of constant diameter.

4.4.1.2 Thermal and Hydrothermal Treatment

Thermal treatment is a drying process accompanied by an increase in temperature, so that the adjustment of both humidity and of temperature is possible. Even more inactivation of undesired enzymes or in general denaturing of proteins are achievable by thermal treatment of the material. Microbiological deterioration is inhibited and the structure of the material is stabilized with respect to the solvent flow during extraction. Furthermore high and short temperature impact–high temperature–short time (HTST) – conditioning is of increasing importance because valuable substances are mobilized without thermal damage.

Heat transfer is strongly enhanced by contacting the natural material with superheated or saturated water vapor. Moreover such treatment in expander machines result in porous structure of the solid which facilitates the fluid flow of the solvent and even the diffusion of the substances to be extracted in the subsequent extraction process [39]. Controlling the condensation of the steam-defined levels of humidity can be adjusted in the natural material that may be of advantage for extractability of polar substances. Sometimes a swelling effect by wetting is desired in order to improve the mass transfer within the swollen pore structure. This method has been used successfully for swelling of raw coffee beans during the decaffeinating process with supercritical CO_2 [40].

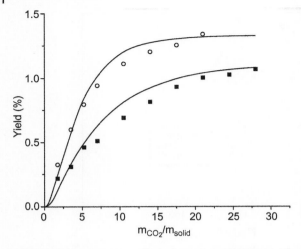

Figure 4.11 CO_2 extraction of valerian, 15 MPa, 50 °C (■ – without pre-treatment, ○ – with CO_2 conditioning).

4.4.1.3 Innovative Methods

Recent developments for the pre-treatment of natural materials have not been introduced in industrial scale so far. Promising methods are the following:

- Enzymatic treatment [34].
- Cell rupture by very fast pressure release. Applicability seems to be advantageous for high value pharmaceutical products, because homogenous cell rupture is restricted to small volumes.
- Swelling by impregnation with CO_2 under elevated pressure. Figure 4.11 illustrates the comparison of supercritical extraction profiles valid for the material valerian with and without swelling before extraction [36].

4.4.2
Extraction

Production of natural substances by supercritical extraction mainly requires process units for solid feed material that has been pre-treated properly. Besides some large scale plants used for industrial production of decaffeinated coffee and tea, and hops extracts, there are a number of smaller multipurpose plants that obtain extracts from a variety of natural materials such as spices, herbs and valuable vegetable oils, flavor compounds or plant pigments like carotenes, aromas and antioxidants. Related to the extraction of natural substances the main advantages of processing with $scCO_2$ are the following: it is inexpensive, not flammable, antibacterial, physiologically harmless; processing is possible at low temperatures, products and residues are free of remaining solvents, the solvating capacity can

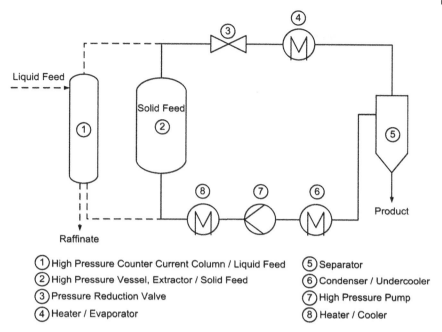

Figure 4.12 General flow sheet for supercritical extraction of solid or liquid feed material.

be enhanced by modifying agents, selective fractionated extraction is possible and finally the scCO$_2$ is recycled within the plant. There are no environmental problems as in processes with conventional solvents. Moreover even by venting the CO$_2$ after the extraction process no CO$_2$ is produced during the process.

Although the range of application for nutraceutical and natural products is still increasing [41] the general process for supercritical processing of the corresponding feed material is a simple cycle consisting of an extraction step and a separation step, combined by a pressurizing pump, for example, a piston pump or even a membrane pump behind, and a controlled pressure reduction valve in front of, the separator. When processing solid material high pressure vessels equipped with quick closures and inner baskets are used and the process is operated batchwise. If the feed material is a liquid, like an essential or vegetable oil, a counter current column will be continuously operated. In order to increase the contacting area for mass transfer these columns are normally equipped with structured packing or filling bodies. Besides these main components heat exchangers are needed for heating, evaporating and condensation of the circulating CO$_2$. Figure 4.12 visualizes the general flow sheet for supercritical extraction of solid and liquid feed material. The process steps are the following: after filling the basket of the pressure vessel with pre-treated phyto extract containing natural material the vessel closures are locked. Liquid CO$_2$ from a storage tank is compressed and heated up to the desired process conditions and it passes through the extraction vessel normally upwards from the bottom to the top. Depending on the specified conditions and corresponding solubility mass transfer of the solutes into the fluid solvent takes place. The loaded supercritical fluid is conveyed to the separation vessel, where the phase equilibrium is changed by adjusting the temperature and the pressure in order to precipitate the solutes. The CO_2 exits the separator in gaseous state. It becomes liquefied in a subsequent condenser before it is recirculated by a pressurizing pump.

As for the range of process parameters, the knowledge of phase equilibrium between the desired extract component and the supercritical fluid is decisive. In the case of extraction of natural components these data often are not available in detail because the desired substances are part of a complex natural structure and other substances are also soluble in the supercritical fluid sometimes even acting as modifying agents. Thus it is of advantage that extraction normally will need small process plants working at pressures between 10 MPa up to 100 MPa, whereas extraction temperatures are sufficient at a limit of 100 °C due to the thermal sensitivity of the products. The volumes of the high pressure vessels vary from 1 L up to 300 L. An important figure evaluating the economics of the process is the total mass of supercritical fluid needed for the extraction of 1 kg feed material. Further, a useful parameter is the relation between the circulating mass flow of the supercritical fluid and the mass of natural material filled in the vessel basket.

4.4.3
Separation

Following the flow schema in Figure 4.12 separation is normally carried out via pressure reduction down to about 6 MPa. Thus the separation is carried out by adjusting a reduced solvent power. Alternative methods of solvent recovery and extract separation also exist for a complete isobaric solvent cycle. The dissolved substances can either be separated in an additional high pressure column by means of an absorbing liquid, for example, water, or can be adsorbed on a fixed bed like activated carbon or silica gel. Moreover separation by adsorption reveals possibilities for creating innovative compound products. As an example antioxidants may be adsorbed on food ingredients like starch. Further, membranes have been proposed [42] for separation of extract and solvent. Due to the relatively low molecular weight of the solvent it permeates the membrane while the extract is recovered. A well established method for separating liquid mixtures is the high pressure spraying with supercritical gas which has solvent power for only one component of the mixture. An example for an antisolvent process is the de-oiling of raw lecithin [43].

Besides actual extraction processes some processes have recently been introduced that use the high solvent capacity of liquids for CO_2. For instance the phase equilibrium of edible fats in contact with $scCO_2$ shows contents of about 30% CO_2 in the oil phase at pressures in the range between 15 MPa and 25 MPa. Those mixtures elucidate low viscosities with excellent spraying characteristics. other processes that have recently shown value are particle generation from saturated solutions (PGSS) and even impregnation of sprayed particles [44].

4.4.4
Post-Treatment

After extraction ultrapure products are often received and as such the target substances might then behave differently compared with the original matrix or composition. For example, the solubility in certain solute might have decreased drastically as natural co-solvents are no longer or only weakly present. While during extraction the substances are dissolved in $scCO_2$ the consistency of the extracts at ambient conditions varies from liquid to oil, waxes and even solids. In many cases this is strongly temperature dependent. Besides phase and rheological behavior the purity might also vary the chemical stability, color appearance, odor etc. All this affects the application of the extracts and they often undergo further process steps after extraction. A few typical post-treatments are described briefly:

4.4.4.1 Degassing
The aim is to remove the dissolved extraction gases for several reasons:

a) **Safety:** special issue if flammable working media are used, for example, propane. However, it is also important to reduce significantly the amount of expanding gases before filling into the product containment.

b) **Appearance:** with even very low (and harmless) gas content the extract might look turbid.

c) **Application:** disturbing substances dissolved gases interfere with final application.

The degassing can be realized by long lasting natural diffusion and evaporation or be supported by moderate heating, steering and supersonic waves.

4.4.4.2 Further Separation
Several standard unit operations are used for further removal of by-products. Adsorption is a common technology for discoloration. Another typical step is to include precipitation by cooling followed by a filtration. In general, all standard unit operations can be applied – although elevated viscosity of the extract can be limiting.

4.4.4.3 Stabilization
As with extraction or fractionation the removal of by-products may have a severe impact on the stability of the final product, especially if the removed components are controlling reactions, this might ease self-modification of the product. Such a removal can be on purpose, for example, as for discoloration, or incidentally but has to be considered in all cases. The addition of stabilizers might be required. For natural products synthetic additives are often no option; however, many natural stabilizers are available, for example, extracts of rosemary are well known for their antioxidant effect.

4.4.4.4 Formulation

Often, the extracts are formulated either for rheological or other application reasons, as is for longer shelf life, or as part of the product specification. Again, polarity is a key issue since the retrieved extracts are no longer in their aqueous or lipophilic environment. In many cases, the extracts are diluted in ethanol, which improves stability but also enhances application into many different systems. Another topic is the removal of remaining water to allow applications in hydrophobic systems.

4.5
Process Design and Operation

4.5.1
Process Concept

Figure 4.13 shows a simplified scheme of a high pressure extraction process. Historically, the extraction from a solid matrix in semi-batch mode is typical for high pressure extraction and still the most used set-up. However, in this figure the scheme also contains a system for feeding liquid raw materials. The column shown on the left replaces the extraction vessel in the case of liquid feed. Usually the $scCO_2$ columns have a lower L/d ratio than for conventional solvents for reasons explained later. Sometimes the batch vessel is taken as extraction unit even for the liquid systems and no column is installed. Yet, since the liquid feed operation is rather close to conventional solvent extraction and as it is less widespread the following will focus on high pressure batch extraction from solids and on $scCO_2$ as solvent.

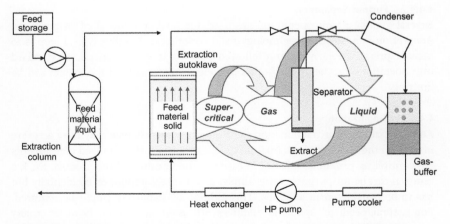

Figure 4.13 Simplified scheme of a typical supercritical fluid extraction from solid matrix, in addition with a liquid feed system (Haneke et al. [45]; original picture from Heidlas [46]).

The process cycle starts with a buffer tank which contains liquid CO_2. From here it first passes a pump cooler to ensure a complete liquid phase to avoid cavitation in the system pump. It pressurizes the CO_2 to target extraction pressure and feeds the extractor. As it passes the heat exchanger the CO_2 reaches the supercritical region. The scCO_2 flows through the solid matrix. The extraction temperature is adjusted by both the heat exchanger before the extractor and a heater outside or inside the extractor vessel. The desired components dissolve in scCO_2 and the mixture leaves the extractor. When the loaded CO_2 passes the back pressure regulator it is depressurized to an intermediate pressure and becomes gaseous. As a result the solubility for the extracted substances drops significantly to almost zero. The pressure is chosen such that the CO_2 does release the extracted components but not too low since the energy demand for the repressurization should be kept low. The released extracts are collected at the bottom of the extractor. As CO_2 exhibits a rather strong Joule–Thomson effect the separator is heated to avoid too low temperatures and to enhance vaporization of the CO_2. Having left the separator, the rather pure CO_2 is cooled to liquefy it; when it re-enters the buffer tank the closed loop starts again.

To recover the extracted feed material from the extractor the CO_2 pump must be stopped, the CO_2 kept in the extractor is smoothly released into the separator until the pressure levels are balanced. Then the extractor is depressurized and the CO_2 is released to atmosphere. Finally, the solids can be removed and new material can be placed in. To reduce batch cycle time and to reduce CO_2 losses high pressure extractions often operate with multiple extraction units in parallel regarding the extraction and sometimes in series regarding the CO_2 flow. This enables a semi-continuous operation. Three parallel extractors are typical. In large scale plants the recovery of the extract is also an issue – this is discussed later. Often several substances can be extracted from one source and it can be attractive to gain them separately. One option is to run the extraction with different extraction parameters (pressure, temperature, solvent-to-feed ratio) several times or to operate a multiple-stage separation, for example, two separators at two different pressure levels or even a fractionation. Some more details will be given later. In general, for the typical semi-batch extraction from solid feed two issues are of major importance: changing pressure level and recycling the CO_2.

4.5.2
Pressurization and Depressurization

One central challenge when operating a high pressure extraction with solid material is pressurizing the autoclave to supercritical pressure in order to run the extraction and later to release the pressure to remove the remaining material. The batch cycle starts with filling the autoclave(s) with the solid feed before it is filled to extraction pressure with CO_2. Due to the compression heat the fluid and the solids are heated up. Also the inside of the autoclave is heated up quickly leading to enormous tension. This has to be taken into account for the detailed design of the autoclaves. When starting the continuous flow of the CO_2 the temperature changes are less dramatic. So, reaching the final extraction temperature can be realized in a more moderate way by increasing the pressure in several steps and allowing flow for some time at intermediate pressures or allowing heat exchange for equilibrating temperature. After extraction the CO_2 feed stream is stopped and the CO_2 is passed to the separator at intermediate pressure (40 . . . 60 bar). In contrast to laboratory or pilot plants the CO_2 remaining in the autoclave is not released to environment. Instead, this volume is sucked with a compressor into the separator. When transferring the CO_2 the pressure decreases and hence the specific volume increases dramatically. As a result the compressor is of remarkable size. For economical reasons the recovery of all CO_2 is prohibitive. At a pressure level ranging from 3 . . . 7 bar the remaining CO_2 is released to the environment. Again, the Joule–Thomson effect induces a strong temperature change following the pressure change. This time the autoclave inside could theoretically cool down to $-78°C$. Subsequently, very high tension within the autoclave material would damage it. Hence, also depressurization is performed in intervals.

4.5.3
CO_2 Recycle Loop

In large scale plants it is economically essential to recycle the CO_2 in a closed loop instead of releasing it to environment as often done in small or pilot plants. Two approaches can be followed: the isobaric mode and operation with intermediate depressurization.

In isobaric mode the CO_2 is cycled at operational pressure and the extract is separated by other means than solubility change of CO_2. One very prominent example is the use of adsorber beds, for example, in tea and coffee decaffeination. The adsorbing medium can be placed in both ways either after the extraction or alternating with the raw material beds. The removal of the extract is often done thermally. As then the extract is destroyed this procedure is only appropriate if the extracts are an undesired side component. Hence, the design of decaffeination plants moved to other separation options since the caffeine can be sold to the beverage industry and others. The selection of the adsorbing medium is driven by the substances to be separated. Usually activated carbon is taken. However, besides capacity, selectivity and separation kinetics, aspects like pressure drop and potential chemical damage to the autoclave are important. Most adsorbents are in pellet form of selected size so that pressure drop is low and well controlled. Several minor components in the adsorbents can attack the autoclave material and lead to corrosion. This has to be checked carefully in advance as the corrosion progress is often very slow but ends with a sudden failure. For food application the adsorbent must be at food grade and is often based on natural substances like fruit kernels or coconut shells.

Ife the extracts have to be recovered, washing towers are used. The added washing medium works as an antisolvent in CO_2, often water or water-based mixtures are used. With such options the extracted components can be sold, for example, caffeine from tea or coffee decaffeination, and costs for adsorbing material and energy are saved. A main difference between adsorber and washing options is that after a batch cycle not only is the extractor with the raw material depressurized, but also the volume of the adsorbent bed, while for the washing tower option the flow rate of the washing medium is continuously brought up to system pressure and released afterwards. A general trade-off between these two options in terms of effort and energy is not possible instead it depends strongly on the phase behavior and the required separation efficiency.

Membrane separation is sometimes discussed as it does not require energy, further fluids and depressurization only for product removal. A scientific study deals with the applicability of different membrane systems in supercritical processing, [47]. but industrial application of this separation option is not known.

The isobaric mode is attractive when rather small fractions of the raw material are extracted. For larger quantities the process has to be run with intermediate depressurization. For identical flow distributions this option requires higher investments and higher operation costs compared with the isobaric option. As described above, the pressure is brought down to such a level that the solubility drops drastically to almost zero. On the other hand the intermediate pressure in the separator is kept rather close below the critical pressure so that repressurization energy is minimized, for example, to 60 bar. The extract is released to a collecting vessel under ambient pressure. Due to the pressure difference of at least 50 bar the extract becomes sprayed and finely dispersed. This can be a safety issue if the extract is flammable (high alcohol content) or a health issue if the extract is toxic or allergenic.

4.5.4 Scale-Up

Key numbers for designing an industrial plant are extraction conditions as pressure, temperature, solvent-to-feed (S/F) ratio and concentration of co-or antisolvent. Furthermore, the solid material defines some parameters by its porosity and particle size distribution as this influences maximum bed height by pressure drop as well as fluid velocity. Besides extraction yield the separation efficiency is also important and is determined by separation temperature and pressure. Along with the process design cost studies should be carried out. Often economic aspects and not thermodynamic determine the process conditions.

The classical way is to scale up a process from laboratory scale trials. However, very often those proposals suffer from severe conceptual failings. Table 4.4 illustrates some typical examples.

Although higher pressures can lead to higher extraction rates still most plants in production are limited to numbers below 500 bar due to earlier limits in approving such plants. In fact, for new plants the pressure level is higher. Furthermore, temperature is at too high level for natural materials. In laboratory systems the residence time in a heated autoclave might be short enough to avoid damages. But in industrial plants the material remains much longer under elevated temperatures than extraction time. Since solubility of $scCO_2$ is rather low, the S/F ratios are high compared with conventional extraction. However, there are limits for S/F

Table 4.4 Comparative values for scale-up of a production process.

HP Extraction	Data from Literature	In production	Reason
Pressure	800 bar	200–500 bar	Max. operating pressure of plant, approval
Temperature	120 °C	Up to 80 °C	Damage of natural materials
Solvent-to-feed ratio	400 kg CO_2/kg feed	Max. 60–300 kg CO_2/kg feed	Productivity
Pressure drop	Often not specified	Max. 1 bar	Limitation of solvent throughput
Co-solvent	30% ethanol	Max. 15–20% co-solvent	Safety, handling and recycling of ethanol. Either CO2 or ethanol

Table 4.5 Key parameters.

	Solubility	Selectivity	Viscosity	Mass transfer	Density difference, kg/m³	Residence time	Max cross flow load
Organic solvent	High	Low	High	Low	100…200	High	Low
Supercritical carbon dioxide	Low	High	Low	High	100…300	Moderate	Moderate

to stay in a reasonable window of productivity. Often the pressure drop is not considered at all but it can be hindering the application, at least its economic attractiveness. Finally, the addition of co-solvents is often stated too high so that safety issues become the driving factor of the plant design. Anyway, the decision to run either a conventional extraction or an SCF extraction has to be made – an addition of more than 20 wt.% of ethanol is not usually reasonable.

For the design of high pressure extraction units and plants there is standard literature available, for example [40, 48]. However, to understand the basic differences between SCF extraction and conventional solvent extraction Table 4.5 lists the key physical properties and some resulting properties adapted from [49].

As an example a continuous extraction from an aqueous raw material stream is considered. For the SCF system the low viscosity and high diffusion coefficient have strong advantages compared with the conventional extraction. While its selectivity is one major criterion for selecting it, the rather poor loading capacity is a major drawback since this leads to the high solvent demands. Pressure differences to penetrate the feed material are for both in the same order of magnitude. Along with the physical data the standard extraction exhibits long residence times but low cross section load, while the SCF extraction operates with moderate residence times and better cross section loads. These theoretical observations lead to the result that the extraction with organic solvents is limited by mass transport. Hence, efforts are put into its optimization, for example, by rotating internals or pulsation. In the case of the $scCO_2$ extraction the limitation is definitely not given by mass transfer but by the loading capacity. Hence, study of the individual phase equilibrium is the key to success.

4.5.5
Costs

Costs are usually stated per kg solid feed material. For the removal of undesired components these are costs for upgrading the material. If the extracts are the targeted substances, the product costs strongly depend on the content in the raw material and the extraction efficiency.

In general, separation costs are mainly driven by throughput of the plant and the number of autoclaves. Although high pressure requires thicker walls and more expensive materials the investment costs of an $scCO_2$ plant are balanced by fewer process units and less effort in downstreaming. Demand for energy is also not a typical criterion when considering the installation of an $scCO_2$ plant. Yet, the operating costs have strong influence as the batch mode requires rather intensive labor work. Roughly stated, the extraction cost for industrial plants ranges from 1 to 5 € kg^{-1} solid, for high throughputs the costs can fall below 0.5 € kg^{-1}. A general study comparing conventional to SCF extraction including detailed cost analysis is given by [50].

4.6
Applications

Supercritical carbon dioxide enables high selectivity even for minor components that are available as well in vegetable oils and other natural extracts. Today, certain minor components are supposed to be of high health value. Often natural extracts from plant material have been produced by conventional methods like solid–liquid extraction or mechanical pressing. As an example relatively high amounts of phytosterols can be found in corn germ oils and rapeseed oils (see Table 4.6) [51]. These minor components have the ability of lowering blood cholesterol and preventing coronary heart disease [52].

Another group of secondary or minor components are vitamins and antioxidative substances that are trace elements in seed oils. Well-known examples are tocopherols, carotenoids and polyphenols. Recently the interest in separating xanthomumols from hop extracts increased strongly due to their assumed anticancer potential [53].

Table 4.6 Phytosterol content in vegetable oil.

Sterol Content (mg/100 g Oil)

Oil	Phytosterol
Corn germ	1000
Cottonseed	324
Olive	221
Rapeseed	821
Soybean	250
Rice germ	1190
Sunflower	340

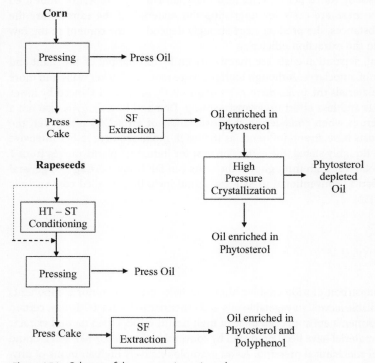

Figure 4.14 Scheme of the process investigated.

Several processes of mobilizing phytosterol components within the plant material in order to enrich them in the extract have been studied. As an example Figure 4.14 shows a flow scheme of rapeseed HTST conditioning followed by mechanical pressing and supercritical extraction of the press cake material. This process chain leads to elevated concentrations of both polyphenols and phytosterols in the extracted oil phase [54].

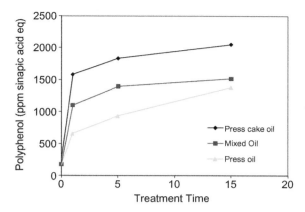

Figure 4.15 Polyphenol content of extract, mixed and press oil samples.

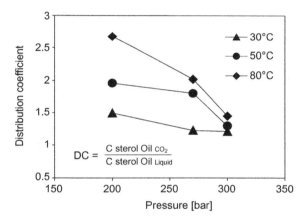

Figure 4.16 Distribution coefficient vs. pressure for initial phytosterol concentration 3%.

By applying the high temperature short conditioning treatment to rapeseeds, not only the normal nutritional effects like deactivating of undesirable enzymes, are accomplished but also the oxidative stability of the oil was increased. The reason is a significantly higher content of polyphenols in the oil. Especially oil extracted from press cake is enriched in antioxidative polyphenols as Figure 4.15 demonstrates.

In comparison oils from normal conditioned seeds elucidate polyphenol contents of only 400 ppm up to 600 ppm. Aiming for high contents of phytosterols in both corn germ and rapeseed oil supercritical extraction was carried out after HTST conditioning. Taking into account the distribution coefficients of phytosterols between the liquid oil phase and a contacting $scCO_2$ phase a further increase in phytosterols could be expected, because values higher than 1 resulted from phase equilibrium investigations [55, 56]. From Figure 4.16 a supercritical process

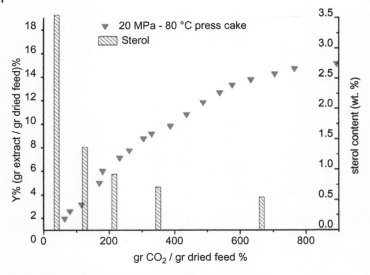

Figure 4.17 Supercritical fluid extraction of corn germ oil.

at pressure and temperature conditions of 20 MPa and 80 °C show a possible enrichment in sterols by a factor of almost 3.

Furthermore it is known that the distribution coefficient normally depends not only on pressure and temperature of the supercritical fluid but also on the concentration of a minor component to be enriched by supercritical extraction. The higher the concentration of a minor component in oil, the higher is the solubility of this phytoextract within the supercritical gas phase. In consequence supercritical extraction of HTST conditioned oilseed leads to a fractionation of the produced oil: the first fraction reveals the highest content of phytosterols. As an example Figure 4.17 demonstrates the high enrichment in sterol content when extracting HTST preconditioned press cake with $scCO_2$ at 20 MPa and 80 °C.

Following common industrial procedures, oil-bearing material, like corn germs and rapeseed, are pre-de-oiled by cage screw pressing. In comparison the oil produced by the extraction of the press residues are generally higher in sterol content than the press oil. As an example Table 4.7 gives corresponding data for the sterol content of oil from untreated and treated rapeseed.

The application of tocopherol enrichment in vegetable oil from wheat germs behaves in a similar way to the fractionated enrichment of phytosterols by supercritical extraction of seed oils. Again a press cake is extracted under conditions with advantageous distribution coefficients. As an example Figure 4.18 gives high contents of the sum of different tocopherols within the first fractions produced by extraction with $scCO_2$ at conditions of 20 MPa and 40 °C [54].

Another prospective application of phytoextraction is revealed by investigations into the production of xanthohumol enriched hop products. Recent pharmacological studies show positive aspects for xanthohumol and related prenylflavonoids of

Table 4.7 Sterol content of oil from untreated and treated rapeseeds.

Treatment	Material	Moisture [wt%]	Oil Content %	Sterol Content %
No Treatment	Mixed oil	7.4	47.43	0.77
	Press oil	–	–	0.63
	Presscake/press oil	–	19.57	0.98
HTST (5 min 145 °C)	Mixed oil	6.3	–	0.8
	Press oil	–	–	0.64
	Press cake oil	6.8	20.97	0.86

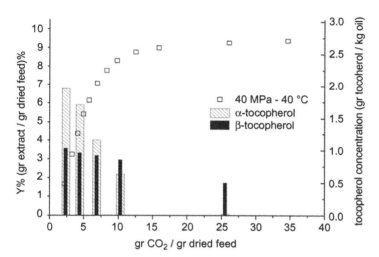

Figure 4.18 Tocopherol fractionation 40 MPa and 40 °C.

hops, for example, possible prevention of osteoporosis or arteriosclerosis as well as anticancer activity [53]. The structural formula of xanthohumol is given in Figure 4.19 [57, 59]. The content of xanthohumol in different varieties of hop ranges from 0.1% up to 1.1% [58], (Table 4.8).

Hop extracts are mainly used as bittering agents in the brewing industry. They are produced by batch extraction processes using $scCO_2$ or liquid ethanol as the solvent. The component xanthohumol has been found to be nearly insoluble in $scCO_2$ whereas this hop ingredient is almost quantitatively extracted by ethanol [58]. The big difference in solubility enables a combined process: a primary hop extraction with ethanol followed by fractionation of the pure resin extract with $scCO_2$. A problem in processing the ethanol produced resin extract is its high viscosity. Moreover, the residue of supercritical treatment of the ethanol extract consisting the enriched xanthohumol ingredients within the insoluble hard resin

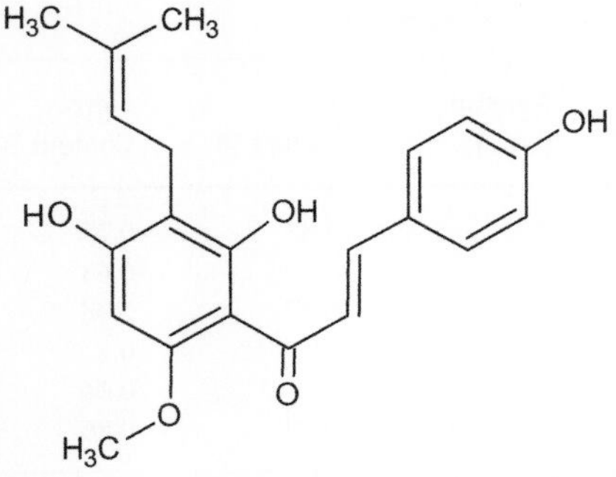

Figure 4.19 Structural formula of xanthohumol.

Table 4.8 Xanthohumol contents of several "Hallertauer" varieties of the crop year 1998 and 1999 [3].

Variety: hallertauer	Xanthohumol content		Ratio xanthohumol to alpha acids	
	crop'98	crop'99	crop'98	crop'99
Taurus	1.0%	0.99%	0.067%	0.065%
Magnum	0.50%	0.42%	0.034%	0.033%
Nugget	0.73%	0.70%	0.066%	0.067%
North.Brewer	0.71%	0.59%	0.069%	0.066%
Perle	0.48%	0.47%	0.079%	0.073%
Tradition	0.38%	0.37%	0.067%	0.063%
Spalter Select	0.42%	0.39%	0.082%	0.088%
Mittelfrüher	0.27%	0.25%	0.066%	0.065%
Hersbrucker	0.22%	0.20%	0.069%	0.083%

fraction elucidates as a high viscous non-fluid almost powder-like product. In order to overcome the corresponding problems of nonhomogeneous and incomplete extraction the high pressure spray process has been developed. Originally this technique was introduced for the de-oiling of high viscous raw lecithin from soy oil refining as an alternative for extraction with acetone [59–61]. The process is shown in Figure 4.20.

The advantage of the process is the mixing zone. A twin nozzle enables an intense contact between the viscous phytoextract and the scCO$_2$ and in consequence the kinetic energy of the high turbulent flow of CO$_2$ leads to a disintegration of the viscous extract into fine droplets with high mass transfer area. Moreover the scCO$_2$ dissolves into the extract and decreases strongly its viscosity. Hence the diffusive mass transfer within the droplets is accelerated. The resulting two-phase

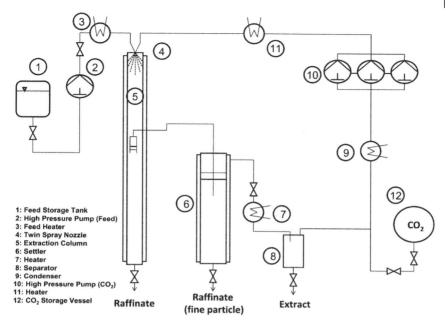

Figure 4.20 High pressure spray process.

1: Feed Storage Tank
2: High Pressure Pump (Feed)
3: Feed Heater
4: Twin Spray Nozzle
5: Extraction Column
6: Settler
7: Heater
8: Separator
9: Condenser
10: High Pressure Pump (CO_2)
11: Heater
12: CO_2 Storage Vessel

flow runs through an extraction zone in which the extraction at very short diffusion paths occurs. Whereas the soluble components precipitate in a separator stage the extract ingredients are gathered at the bottom of the extraction unit which is normally constructed as a high pressure column. This process is generally applicable for those extracts being as valuable ingredients in products of conventional solvent extracts but remaining as the residue after a subsequent treatment with $scCO_2$.

Thus a similar process has been developed for extracts that are ingredients in by-products of the edible oil industry. Liquid feed material like carotene-containing dispersions and oleoresins are well-examined sources for products with enriched levels of carotenes. Triglycerides show solubilities in $scCO_2$ that exceed the data for carotenes (beta-carotene, lutein, astaxanthin, lycopin) by three orders of magnitude [62, 63]. Aiming at products with enriched contents of carotenes the supercritical spray process has been introduced using the high distribution coefficients between carotenes and triglycerides in $scCO_2$. Carotene-containing dispersions or even palm oil is sprayed into a high pressure column. The oil is dissolved by the $scCO_2$ and the residue elucidates high concentrations of carotenes.

An example is the enrichment of beta-carotenes from by products of the palm oil industry. The high pressure spray process results in a refinement of beta-carotenes up to concentrations of nearly 80% [63]. A further application is the separation of carotenoids from solvent extracted paprika oil. Other promising applications of natural products extraction with supercritical fluids will be developed in future.

Acknowledgments

The authors wish to thank Dr. E. Schütz for supporting this chapter. His contribution exceeded the provision of published data as he updated data solely for this book. Furthermore, his consultancy helped the authors to weight and reflect the individual issues of supercritical fluid extraction properly.

References

1 Schneider, G.M., Stahl, E. and Wilke, G. (1980) *Extraction with Supercritical Fluids*, Wiley-VCH Verlag GmbH, Weinheim, Germany.
2 Stahl, E., Quirin, K.W. and Gerard, D. (1998) *Dense Gas for Extraction and Refining*, Springer Verlag, Berlin, Heidelberg.
3 Liu, K. and Bork, M. (2001) *Extraction of Natural Substances with Supercritical Carbon Dioxide and Co-Solvent*, The International Symposium GVC, Hamburg, Germany.
4 EEC (1988) EEC: Richtlinie 88/344/EWG des Rates vom 13. Juni 1988 zur Angleichung der Rechtsvorschriften der Mitgliedsstaaten über Extraktionslösemittel, die bei der Herstellung von Lebensmitteln und Lebensmittelzutaten verwendet werden: Amtsblatt Nr. L 157 vom 24/06/1988, S. 0028-0033.
5 Pilz, S., et al. (2006) Überkritische Extraktion – aus Sicht der Industrie, in *Fluidverfahrenstechnik*, (ed. R. Goedecke), Wiley-VCH-Verlag GmbH, Weinheim, Germany, pp. 1074–1130.
6 Schütz, E. (2003). Personal communication, Databank on SCF patents.
7 Clifford, A.A. (1995) Chemical destruction using supercritical water, in *Chemistry of Waste Minimization* (ed. J.H. Clark), Blackie Academic, Glasgow, UK, pp. 504–521.
8 McHugh, M. and Krukonis, V. (1986) *Supercritical Fluid Extraction, Principles and Practice*, Butterworths, Boston.
9 Bruno, T.J. and Ely, J.E. (1991) *Supercritical Fluid Technology, Reviews in Modern Theory and Applications*, CRC Press.
10 Yoshioka, M., Parvez, S., Miyazaki, T. and Parvez, H. (eds) (1989) *Supercritical Fluid Chromatography and Micro-HPLC, Progress in HPLC*, vol. 4, VSP, Utrecht, The Netherlands.
11 Saito, M., Yamauchi, Y. and Okuyama, T. (eds) (1994) *Fractionation by Packed-Column SFC and SFE: Principles and Applications*, VCH, New York.
12 Savage, P.E., et al. (1995) *AIChE J.*, **41**, 7.
13 Jessop, P.G. and Leitner, W. (eds) (1999) *Chemical Synthesis Using Supercritical Fluids*, Wiley-VCH Verlag GmbH, Weinheim, Germany.
14 Bungert, B., Sadowski, G. and Arlt, W. (1998) *Ind. Eng. Chem. Res.*, **37**, 3208.
15 Jung, J. and Perrut, M. (2001) *J. Supercrit. Fluids*, **20**, 179.
16 Bach, E., Cleve, E. and Schollmeyer, E. (1998) in Proceedings of the Fifth Meeting on Supercritical Fluids: Materials and Natural Products Processing (eds M. Perrut and P. Subra), Nice, p. 345.
17 Cooper, A.I. (2000) *J. Mater. Chem.*, **10**, 207.
18 Gmehling, J. and Kolbe, B. (1992) *Thermodynamik*, 2nd edn, VCH.
19 Cagniard de la Tour, C. (1882) *Ann. Chim.*, **22**, 410.
20 Perry, R.H., Green, D.W. and Maloney, J.O. (eds) (1984) *Perry's Chemical Engineering Handbook*, 6th edn, McGraw-Hill.
21 Poliakoff, M., George, M.W. and Howdle, S.M. (1997) Inorganic and related chemical reactions in supercritical fluids, in *Chemistry under Extreme or Non-Classical Conditions* (eds C.D. Hubbard and R. van Eldik), John Wiley & Sons, Inc., New York.

22 Dinjus, E., Fornika, R. and Scholz, M. (1997) Organic chemistry in supercritical fluids, in *Chemistry under Extreme or Non-Classical Conditions* (eds C.D. Hubbard and R. van Eldik), John Wiley & Sons, Inc., New York.

23 Cygnarowicz-Provost, M. (1996) *Design and Economic, Design and Economic Analysis Os Supercritical Fluid Extraction Processes, Supercritical Fluid Technology in Oil and Lipid Chemistry*, AOCS Press, Champaign, Illinois. Ch.7.

24 King, M.B. and Bott, T.R. (1987) *Sep. Sci. Technol.*, **22**, 1103.

25 Li, L. and Kitan, E. (1989) *Supercritical Fluid Science and Technology, ACS Symp*, vol. 406 (eds K.P. Johnston and J.M.L. Penninger), American Chemical Society, Washington, p. 317.

26 Knez, Z., Posel, F., Hunek, J. and Golob, J. (1991) Supercritical fluids II. Proc. Int. Symp., Boston, U.S.A., p. 101.

27 Goto, M., Smith, J.M. and McCoy, B.J.V. (1990) *Ind. Eng. Chem. Res.*, **29**, 282.

28 De Arévalo, M. (2008) Phytosterol enrichment in vegetable oil by high pressure processing, PhD Thesis, Hamburg-Harburg.

29 Incropera, F.P. and De Witt, D.P. (2002) *Fundamentals of Heat and Mass Transfer*, John Wiley & Sons, Inc., USA.

30 King, M.B. and Bott, T.R. (1993) *Extraction of Natural Products Using Near-Critical Solvents*, Blackie Academic, Glasgow, UK.

31 Wakao, N. and Kaguei, S. (1982) *Heat and Mass Transfer in Packed Beds*, Gordon and Breach, New York, vol. 153, p. 139.

32 Wakao, N. and Smith, J.M. (1962) *Chem. Eng. Sci.*, **17**, 825.

33 Eggers, R., Sievers, U. and Stein, W. (1985) *JAOCS*, **62**, 8.

34 Perrut, M. (1994) *Chem. Biochem. Engi. Q*, **8**, 25.

35 Sovová, H., Kučera, J. and Jež, J. (1994) *Chem. Eng. Sci.*, **49**, 415–420.

36 Stamenic, M., Zizovic, I., Eggers, R., Jaeger, P., Heinrich, H., Rój, E., Ivanovic, J. and Skala, D. (2010) *J. Supercrit. Fluid*, **52**, 125–133.

37 Mertens, B. and Knorr, D. (1992) *Food Technol.*, **46**, 124–128.

38 Tzia, C. and Liadakis, G. (2003) *Extraction Optimization in Food Engineering*, Marcel Dekker, New York, p. 126.

39 Schütz, E. (2004) Charakteristika verdichteter Gase bei der Naturstoffextraktion. Pro3-Minisymposium, "Neue Wege und Lösemittel in der Extraktion"; TU Kaiserslautern; 29.09.2004.

40 Bertucco, A. and Vetter, G. (2001) *High Pressure Process Technology*, Elsevier, Amsterdam.

41 Martinez, J.L. (2008) *Supercritical Fluid Extraction of Nutraceuticals and Bioactive Compounds*, CRC Press, USA.

42 Gehrig, M. (1986) Verfahren zur Gewinnung von Coffein aus verflüssigten oder überkritischen Gasen. DE 3443390A1.

43 Wagner, H. and Eggers, R. (1996) *AIChE J.*, **42**, 1901.

44 Wendt, T. (2007) Herstellung flüssigkeitshaltiger pulverförmiger Komposite durch ein Hochdrucksprühverfahren für Anwendungen im Lebensmittelbereich. Dissertation, Bochum.

45 Haneke, M., *et al.* (2009) Extraction of Natural Materials – an Analysis on Influence Factors for (Non-) Commercialization, Presentation at ACHEMA, 15.09.2009; Frankfurt.

46 Heidlas, J.E. (2001), Personal communication.

47 Sartorelli, L. (2001) Abtrennung von Extrakten aus überkritischen Gasen mittels Membranen, PhD Thesis, TU Hamburg-Harburg.

48 Brunner, G. (1994) *Gas Extraction*, Springer, New York.

49 Pilz, S. and Marckmann, H. (2007) Phasenverhalten und Stoffübergang unter Hochdruckbedingungen – Know-How übertragbar in den industriellen Maßstab? DECHEMA-Kolloquium "Prozesse mit überkritischen Fluiden", 01.02.2007, Frankfurt.

50 Pellerin, P. (2003) Comparing extraction by traditional solvents with supercritical extraction from an economic and environmental standpoint. International Symposium on Supercritical Fluids; Versailles, France, April 2003.

51 Gunstone, F. (2001) *Eur. J. Lipid. Sci. Technol.*, **103**, 447.
52 Moreau, R., Whitaker, B. and Hicks, K. (2002) *Prog. Lipid. Res.*, **41**, 457.
53 Stevens, J.F., Mirinda, C.L., Buhler, D.L. and Deinzer, M.L. (1998) *J. Am. Soc. Brew. Chem.*, **56**, 136.
54 Zacchi, P., Daghero, J., Jaeger, P. and Eggers, R. (2006) *Brazilian J. Chem. Eng.*, **23**, 105.
55 Zacchi, P., Michel, A. and Eggers, R. (2007) Supercritical CO_2 processing for phytosterol enrichment in vegetable oils. Iberoamerican Conference on Supercritical Fluids PROSCIBA, 2007.
56 Michel, A. (2008) Phytosterol enrichment in vegetable oil by high pressure processing, PhD Thesis, Hamburg-Harburg.
57 Biendl, M., Eggers, R., Czerwonatis, N. and Mitter, W. (2000) *Lebensmitteltechnik*, **11**, 74–76.
58 Kammhuber, K., Zeidler, C., Seigner, E. and Engelhard, B. (1998) *Brauwelt*, **138**, 1633.
59 Stahl, E. and Quirin, K.-W. (1985) *Anstrichmittel*, **87**, 6, 219.
60 Eggers, R. and Wagner, H. (1993) *J. Supercrit. Fluids*, **6**, 31.
61 Johannsen, M. (1995) Experimentelle Untersuchungen und Korrelierung des Löseverhaltens von Naturstoffen in überkritischem Kohlendioxid, PhD Thesis, TU Hamburg-Harburg.
62 Pietsch, A. (2000) Die Gleichstromversprühung mit überkritischem Kohlendioxid an den Beispielen Hochdruckentcoffeinierung und Carotioidaufbereitung, PhD Thesis, TU Hamburg-Harburg.
63 Eggers, R., Pietsch, A., Lockemann, C.A. and Runge, F. (1999) Processing carotenoid-containing liquids with supercritical CO_2. International Meeting of the GVC-Fachausschuss "Hochdruckverfahrenstechnik", Forschungszentrum Karlsruhe, Germany, March 3–5, 1999.

5
Process Engineering and Mini-Plant Technology

Jochen Strube, Werner Bäcker and Michael Schulte

5.1
Introduction

Process development for plant-based extracts can in general be divided into two different **tasks**:

1) Mixtures or pure components of known efficacy are the objects of the process development task. For example, a new or different plant may be chosen in order to substitute the raw material for a known mixture or pure component as a new product. Analytical results have already been established in advance.

2) Different raw materials are analyzed totally or partially in order to detect ingredients as products in the form of mixtures or pure substances. Analytical results must be developed in parallel to product screening and testing in most cases.

All these approaches need sound and efficient **analytical methods** to determine the target compound or fraction and site key-components in complex mixtures of plant extracts from a solid–liquid extraction (SLE). Target compounds or fractions are within a range of tens to hundreds of compounds where the target has no distinct physical properties like molecular weight/size, isoelectric point, hydrophobicity, solubility etc. and therefore cannot be separated with a single unit operation.

Plant extracts contain **compounds** from different chemical groups, for example, lipids/fats, sugars, organic acids, polyphenols, terpenes, etc. [1]. The mixture of compounds extracted is dominated by the extraction media and operation conditions [2–8]. A general approach to developing the SLE step is given in Figure 5.1, which summarizes the different process concepts and related unit operations to gain different target compounds or mixtures from different sources [9,10].

Process development is very product specific and therefore product chemistry and physical properties must be taken into account. The choice of extraction media is related to the physical properties of the target compound/mixture. Its location

Industrial Scale Natural Products Extraction, First Edition. Edited by Hans-Jörg Bart, Stephan Pilz.
© 2011 Wiley-VCH Verlag GmbH & Co. KGaA. Published 2011 by Wiley-VCH Verlag GmbH & Co. KGaA.

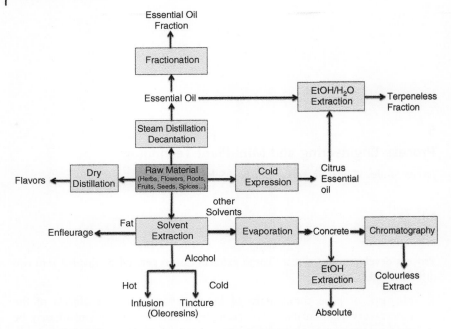

Figure 5.1 General overview of process concepts for the extraction of natural products (Jenelten et al., Firmenich S.A) [9].

within the plant determines the necessary size of plant grinding and the residence time necessary to gain high yields. Extraction media and target component, and also by-product environment within the plant fractions, determine the necessary selectivity to gain a given purity.

To explain the **general guidelines** three main cases can be differentiated:

- If the target component or mixture is polar, for example, hydrophilic, then it can be assumed that the target fraction is homogeneously distributed over the plant, because the plant consists of water-soluble compounds separated into cells by membranes, which are more hydrophobic/lipophilic. Distribution is also assumed to be located throughout the plant. Extraction kinetics is therefore assumed to be medium fast/slow. The plant must be cut into at least medium size fractions. A hydrophilic extraction media should be chosen. Dry plant material tends to need some residence time for opening pores to allow mass transfer.

- If the target component or mixture is apolar for example, hydrophobic, then it is assumed to be surrounded by a more oil-like environment within the plant cells. A hydrophobic extraction media should be chosen. Dry raw material should be treated with polar media in advance to swell the pores again and wet the hydrophilic plant parts. As a consequence extraction kinetics are comparatively slow. Plants sources must be ground into smaller fractions.

- If the target compound or mixture is outside of the plant, located in a fat-/wax-like environment, the extraction kinetics should be fast if an appropriate apolar solvent is used.

Chromatography methods have proven to be the standard workhorse to cope with analytical challenges efficiently [11–15]. In most cases two-dimensional (2D) chromatography is applied. The fractions of a first normal phase (NP) or reverse phase (RP) separation are further resolved by a second orthogonal RP, NP or ion exchange method. Method development has systematic rules which are described in Section 5.2.

After an identification test on the analytical compound, samples must be supplied for screening, efficacy testing or structural analysis. Preparative chromatography has proven to be the most efficient and reliable process in that step, because the necessary separation steps can be predicted from 2D chromatography. Scale-up to preparative chromatography is straightforward and well-known. Preparative chromatography supplies for example, test amounts of pure components in µg- to 5 mg-scale for GC-MS and NMR.

The spin-off from this analysis is to predict a potential separation sequence for the production process to be developed. The direct scale-up of chromatographic methods into production scale is in most cases too expensive, even if it is operated with full solvent recycling. Therefore the next logical step is the development of any **liquid–liquid extraction (LLE)** method to make the separation tasks for chromatography easier and cheaper by fractionating, and for example, generating polar and/or apolar fractions in advance by LLE steps.

Development of potential LLE methods starts with a classical solvent screening. Hints may be obtained by calculating the solubility parameters of compounds by thermodynamic concepts like Hansen, Maurer etc. [16–20]. It is obvious that process development of a LLE step includes deciding whether LLE in **combination** with chromatography, or LLE alone, is the best choice [21].

From this line of argument it is obvious that in process development starting with any combination of LLE and chromatography the SLE operation (that includes the raw material preparation) must also be taken into account. Because, if SLE is started with a water system, then an organic phase is needed for LLE, or vice versa: if any organic solvent is applied for SLE, then a water phase is needed for LLE to gain a miscibility gap for the two liquid phases. The strong interference of SLE and raw material preparation is discussed in Sections 5.1 and 5.6. The extraction media in LLE chosen – water or organic system – has consequences for the choice of the most economic chromatography method, for example, either NP or RP.

In addition, because the solubility of a compound into a solvent media may be incomplete, screening of **unit operation alternatives like crystallization and precipitation** follows, which is in most cases dominated by the formulation task needed. Separation is normally treated as a subtask, but the formulation task dominates. Precipitation of low solubility unwanted impurities like sugars, proteins, fats/lipids is applied as a standard procedure.

Membrane technology is a workhorse which concentrates highly diluted process streams (0.01–1 g L^{-1}) up to the limiting solubility in order to reduce volumetric flow rates and thus downstream equipment size. Therefore, ultra-filtration and nano-filtration are typically applied more often between each main step following the molecular-weight-cut-off (MWCO) to change solvents or cut molecular groups due to their size, like lipids/fats, sugars and proteins (see Table 5.1 and Figure 5.6).

Volatile compounds, like essential oils, are normally produced by hydro (steam) distillation. Distillation, evaporation and condensation is also applied to recycle solvents in process conditions with about 0.1–1% losses per year if economically necessary [23].

Taking into account this description of a systematic approach for process development it can be seen clearly, that as a consequence **total process development** is needed when considering the full process chain, which must be started with the development of appropriate analysis. There are many alternatives so a trial-and-error or solely experimental approach without a sound theoretical basis will consume too much effort, time and resources. Therefore, in this chapter a systematic and efficient process development and design approach will be described.

Today, methods of chemical engineering in process development and conceptual process design include mini-plant technology with process modeling combined with experiments in single mini-plant units to determine the model parameters. Single unit operations are modeled and their model parameters determined at first. Then the single units are combined by total process modeling in any sequence and the most efficient is chosen and proposed in the conceptual design phase. After economic and operation evaluation the best process sequence is experimentally validated in a mini-plant. Sufficient test quantities are supplied by that. Mini-plant technology allows reliable scale-up into pilot or production scale. Detailed engineering may start to design the production-scale plant. Custom manufacturer capacities may be evaluated to supply products fast to the market in parallel or advance.

General concepts for process development of separation sequences for mixtures by chromatography unit operations are described in [2]. It should be kept in mind, that process development will at first propose a process conceptual design which will lead, if thermodynamically feasible, to an early economic evaluation. If this is positive, then experimental verification by operation of a mini-plant is necessary. Before that, detailed experiments on each single unit operation are necessary to determine the operation region and/or model parameters for process optimization by simulation studies. Statistical design of experiments combined with model parameter determination for rigorous (physical-chemical) models are the efficient way [10, 13]. An overview is given in Figure 5.2.

The majority of specific operation costs are fixed in an early process development stage before basic engineering [2, 24], as shown in Figure 5.3. Therefore, an efficient method for **feed characterization** of complex mixtures is needed as a basis of existing process conceptual design methods by process simulation, see Figure 5.3 [22, 25–28]. Multidimensional preparative HPLC chromatography has been

Table 5.1 Main groups and subgroups of plant ingredients [1, 22].

Main groups	Subgroups
Carbohydrates and lipids	Monosaccharides
	Oligosaccharides
	Polysaccharides
	Sugar alcohols and cyclitols
	Organic acids
	Fatty acids and lipids
	Hydrocarbons and derivates
	Acetylenes and thiophenes
	Miscellaneous aliphatics
Nitrogen-containing compounds (excluding alkaloids)	Amino acids
	Amines
	Cyanogenic glycosides
	Glucosinolates
	Purines and pyrimidines
	Protein and peptides
	Miscellaneous nitrogen compounds
Alkaloids	Amaryllidaceae alkaloids
	Betalain alkaloids
	Diterpenoid alkaloids
	Indole alkaloids
	Isoquinoline alkaloids
	Lycopodium alkaloids
	Monoterpenoid and sesquiterpenoid alkaloids
	Peptide alkaloids
	Pyrrolidine and piperidine alkaloids
	Pyrrolizidine alkaloids
	Quinoline alkaloids
	Quinolizidine alkaloids
	Steroidal alkaloids
	Tropane alkaloids
	Miscellaneous alkaloids
Phenolics	Anthocyanins and anthochlors
	Benzofurans
	Chromones and chromenes
	Coumarins
	Minor flavonoids
	Flavones and flavonols
	Isoflavonoids and neoflavonoids
	Lignans
	Phenols and phenolic acids
	Phenolic ketones
	Phenylpropanoids
	Quinones
	Stilbenoids
	Tannins
	Xanthones
	Miscellaneous phenolics

Table 5.1 Continued

Main groups	Subgroups
Terpenoids	Monoterpenoids
	Iridoids
	Sesquiterpenoids
	Sesqiterpene lactones
	Diterpenoids
	Triterpenoid saponins
	Steroid saponins
	Cardenolides and bufadienolides
	Phytosterols
	Cucurbitacins
	Nortriterpenoids
	Miscellaneous triterpenoids
	Carotenoids
	Acetylenes and thiophenes
	Miscellaneous aliphatics

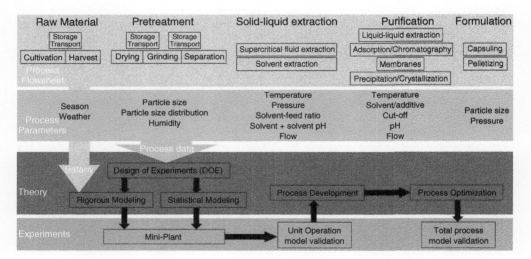

Figure 5.2 Chemical engineering approach for process development [10].

proven to be most efficient to generate the necessary specification amounts for analysis. A visualized workflow is given in Figure 5.4.

Peak purity and spectral libraries via DAD have proven not to be sufficient for multicomponent plant extracts [22]. Therefore, a second HPLC dimension is needed to decide on peak purity and to increase resolution, as shown in Figure 5.5 [22]. Based on that, 2D-HPLC on the preparative scale has proven to be an economic workhorse to solve the problems of efficient sample supply and chemical characterization of many ingredients within complex mixtures. [22].

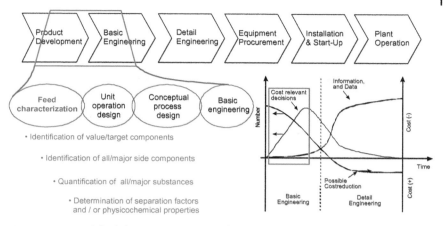

Figure 5.3 General feed characterization method [22, 24].

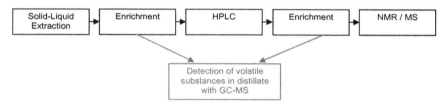

Figure 5.4 Method development for separation and identification of natural products using the example of plant extracts [22].

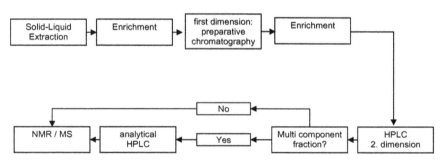

Figure 5.5 Method development for separation and identification of natural products using the example of plant extracts – 2D-HPLC [22].

All systematic approaches to process development of plant-based products should start with the **botanical–chemical basis** to define the target component or mixture by individual structures or groups. Nevertheless, the final objective should always be kept in mind so that process steps can be developed or designed with appropriate operation conditions for the production scale.

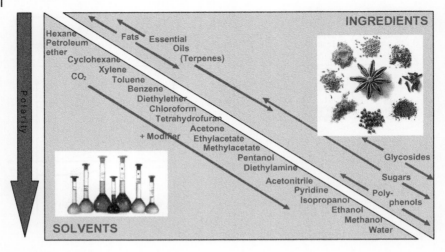

Figure 5.6 Ingredient vs. solvent polarity [10].

In general, ingredients of plants can be sorted into 5 main and 64 subgroups, summarized in Table 5.1. Based on these, the first rough estimates of physical properties of the target compound or mixture can be found, due to mainly polarity and chemical stability. Polarity determines the choice of pre-treatment, SLE, LLE and HPLC solvent type choice (Figure 5.6).

The range of product stability determines pH, enzymatic pre-treatment and temperature at a specific pressure range of operation conditions of all unit operations in combination with proper residence times, in order to gain maximum yield and activity of a target component/mixture. With a given polarity of the target component/mixture a solvent should be chosen, where Figure 5.7 gives information on the polarity of solvents.

In addition to that, only media permitted by regulatory authorities can be chosen, as also discussed in Chapter 1 [29–35]. A further limitation is the approval of the manufacturing side to specific media handling under, for example, explosion-proof, aqueous and environmentally friendly conditions. A summary of **regulatory limitations** of the application of auxiliaries like solvents for each product group, is given, [29, 30, 34].

1) Solvents in pharmaceuticals [33–35]
    - class 2: solvents limited due to a TDI (tolerable daily intake)
        - hexane
        - methanol
    - class 3: solvents with low toxic efficacy (limitation)
        - ethanol
        - heptane
        - 2- propanol.
2) Solvents in food/nutraceuticals etc. [30]

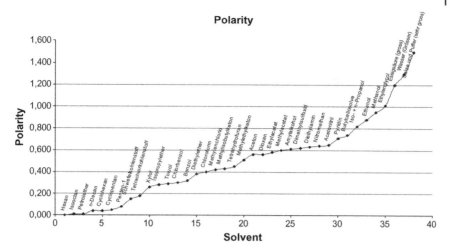

Figure 5.7 Solvent polarity scale [2].

The regulatory basis for the use of solvents is defined in an administrative order (Hilfsstoffverordnung) from which we can extract paragraphs 2, 3, and 4 from the executive summary:

- § 2 Permitted solvents are to be found in amendment 2 as listed components for specific applications defined there:
  - hexane
  - methanol
  - propan-2-ol
- § 3 Moreover, drinking water, ethanol and other foods which have solvent properties, are permitted as extraction media.
- § 4 purity specifications:
  - In amendments 1 to 3 listed components are only permitted for commercial producers of food as extraction media, which are determined to be marketed, if purity specifications following paragraph 4 amendments are fulfilled. The same procedure must be followed concerning use of ethanol.

The following logical decision tree is proposed, based on the botanical position and polarity of the target component/mixture, the hardness of the raw material and its moisture content (Figure 5.8). Position and hardness dictate the required residence time for proper pre-treatment and solid extraction. They also determine the necessary degree of grinding.

If the raw material is dry then water is commonly used to swell the dry material again and open the pores and membranes for diffusion of the extraction solvent

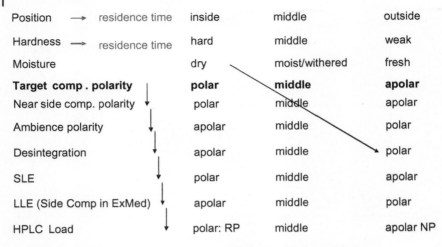

Figure 5.8 Decision tree for (pre-treatment)- SLE–LLE–HPLC solvent choice.

and target component/mixture. This is especially the case if the target component is apolar and consequently an apolar solvent is chosen, which is good for extracting the target component, but is bad for wetting and will not swell the raw material matrix.

Steam distillation for essential oil production uses this approach perfectly. With the addition of water which is heated and evaporated or directly with injected steam for heating, water is condensed in the colder plant matrix and extracts the target component/mixture. Since the matrix is swollen and pores are opened the diffusion out into the surrounding steam is easy, where the solute is evaporated and transported to the top of the column to be re-condensed and sampled [32].

The following scheme in Figure 5.9 gives an overview of necessary steps in the conceptual process design for plant-based extract products. [22].

5.2
Chromatographic Screening

Chromatography methods are the workhorse for the identification of interesting ingredients, which should be purified on the preparative scale for sample testing, efficacy screening and/or structure analysis. One of the biggest advantages of chromatographic separation technologies is the high variability in different selective principles which can be used to isolate or purify target compounds. In addition, chromatography or chromatographic adsorption can be scaled-up directly from experiments, which have been done on the analytical scale (but not necessarily following analytical rules). Therefore the identification of interesting compounds by chromatographic techniques and the subsequent scale-up of the production process can be done in a consistent and straightforward manner.

5.2 Chromatographic Screening

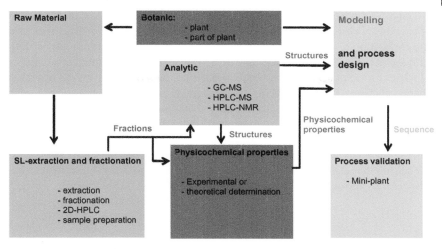

Figure 5.9 Overview of work packages for conceptual process design of plant-based extracts. Work packages which are already state-of-the-art are colored green, the existing conceptual design models ready for use are in yellow and in red are bottlenecks still under research [22].

A simple pre-fractionation procedure uses RP chromatography for the isolation of plant extracts. Especially medium-polar extracts, for example, ethyl acetate, methanol or ethanol extracts can be directly fractionated by RP chromatography. To increase the efficiency of the separation the use of a RP-flash cartridge with a high plate number is advisable. Good results can also be achieved when a monolithic column is used. The Chromolith Prep column offers a high efficiency of up to $20.000\,N\,m^{-1}$ at a pressure drop which allows the operation of the column in a standard flash chromatography setting. Figure 5.11 shows a chromatogram of an ethanolic plant extract, which has been separated using a Büchi Sepacore system together with a Chromolith Prep. An acetonitrile/water gradient has been used rising from 5% acetonitrile to 80% within 25 minutes. Special emphasis must be put on the detection of all compounds. Single-wavelength UV-detection is very limited, diode-array detection (DAD), evaporating light scattering detection (ELSD) or coupled mass spectrometry (MS) also shows compounds which exhibit no adsorption under UV light. Special emphasis must be laid on compounds that are not eluting from the column, because in preparative isolation irreversible adsorption is the reason for column clogging.

A very versatile tool for process development in plant compound isolation is thin layer chromatography (TLC). TLC in its modern set-up can be used in different elution modes and with special detection technologies. One of the biggest advantages of TLC is the possibility to detect all compounds. Especially non-moving compounds at the starting spot of the TLC plate can be easily detected before they cause any trouble in the column. For the group-specific detection of natural compounds a variety of different spraying solutions have been developed [36].

Today a lot of research is done to couple TLC to MS detection. One of the first approaches was reported by Van Berkel who developed a TLC-coupling device for the connection of a desorption electrospray ionization mass spectrometry (DESI-MS) system [37, 38]. A sampling capillary (550 µm i.d.) was mounted in a distance of 50 µm on top of a TLC plate. The TLC plate can be scanned by means of a x,y,z-robotic platform. Using this set-up alkaloids in six different over-the-counter goldenseal dietary supplements from *Hydrastis Canadensis* have been examined qualitatively and quantitatively. A standard curve has been set up for the three alkaloids berberine, palmatine and hydrastimine. The TLC separation condition for the alkaloids was with normal phase silica gel chromatography using ethyl acetate/methanol/formic acid/water as the mobile phase. In addition Bruker has just recently introduced a TLC-coupling device for a MALDI-MS system.

5.3
Preparative Task

As an example of a preparative HPLC concept the isolation of compounds from an extract of *Juglans regia* is used. In the Middle East countries a tea preparation of the dried kernels from the Persian walnut (*J. regia*), (Figure 5.10) is used to treat diabetes. Typically, insulin sensitizers are applied like glitazones or glinides [40]. The isolation of the main components from *J. regia* is used in the following to describe the principle of chromatographic method development and scale-up in the isolation of compounds from plant material.

The dried walnut kernel is ground in a mill and suspended in a 500 mL round flask with either water, ethanol, ethyl acetate or *n*-hexane. After 24 h the solvents are heated up to reflux for 8 h. After cooling down the suspensions are filtered off and the solvents removed. The different extraction solvents yield different dry masses:

water	1050 mg
ethanol	800 mg
ethyl acetate	240 mg
n-hexane	250 mg

The four different extracts are analyzed using RP-HPLC with diode-array detection using an acetonitrile/NaH$_2$PO$_4$-buffer (pH 2.6) gradient on a Chromolith® Performance RP-18e. Figure 5.11 shows the corresponding chromatogram of the ethanol extract. The two main compounds elute at 4.76 min (P3) and 5.96 min (P4) and should be further evaluated, as their amount in the chromatogram seems to make them good candidates for a preparative isolation. The DAD spectra for P4 could correspond to flavonoid structures of the eriocitrin or taxifolin type while the spectrum for P3 could be from an epicatechin or procyanidin compound.

The further peak identification is done by TLC using specific detection reagents. For TLC identification three different solvent systems are used, the details for the most common are given below.

Figure 5.10 Persian walnut (*Juglans regia*) [39].

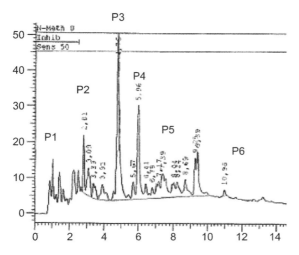

Figure 5.11 Chromatogram of ethanolic walnut extract.

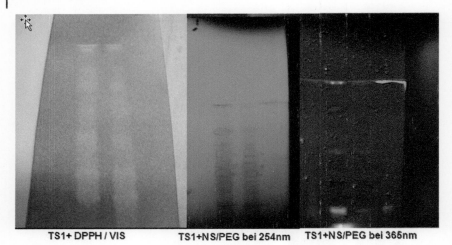

TS1+ DPPH / VIS   TS1+NS/PEG bei 254nm   TS1+NS/PEG bei 365nm

Figure 5.12 TLC-chromatogram of ethanolic walnut extract.

- TLC-plates: Silica gel 60 F254 (20×20 cm, active height 0,5 mm, Fa. Merck, Art. 1.05715)
- Mobile phase: Ethyl acetate–glac. Acetic acid–formic acid–water (100+11+11+27)
- Detection: DPPH: anti oxidant quick test
- NS/PEG: natural compound/polyethylene glycol at 254 and 365 nm

The TLC chromatograms of the ethanolic extract are shown in Figure 5.12. The ingredients of the peaks are controlled by MS-detection, then an inspection of the corresponding peaks was taken and it was decided to isolate the peaks P3 and P4 from the ethanolic extract.

For the further evaluation of preparative chromatographic isolation strategies the peak groups from the original chromatogram are symbolized by differently colored triangles in Figure 5.13, which will be used to show the retention behavior of the different compounds and compound groups. In general the peaks can be characterized as:

P1 – non-binding compounds, eluting in flow-through
P2 – early eluting compounds, limited adsorption on the chromatographic phase
P3 – main compound
P4 – close relating impurity, main separation task (in our case the selectivity between P3 and P4 in the original chromatogram is $\alpha=1.31$)
P5 – late-eluting compounds, higher affinity to the stationary phase
P6 – sticky compounds, strongly binding to the stationary phase.

One of the main mistakes in preparative chromatography is the submission of a crude fraction towards the stationary phase. If in our case the whole mixture of P1–P6 is submitted to the phase several through-put limiting phenomena will occur:

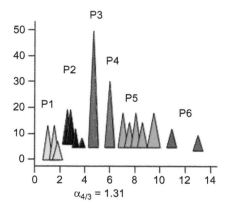

Figure 5.13 Reduction of chromatograms of ethanolic walnut extract.

- When increasing the load on the column the linear range of the adsorption isotherm will very soon be left and strong interaction phenomena between the different compounds will take place. This will limit the possible yield at a given purity for the main compound. As the selectivity between our main compound (P3) and the closely related impurity (P4) is rather small ($\alpha = 1.31$) the productivity for a preparative chromatography is low.

- The second phenomenon occurring with a mixture of P1–P6 is the problem of poisoning the column with the sticky compounds P6 and partly P5. Especially in stainless steel columns it is often problematic to see the amount of strongly binding compounds on the adsorbent. These compounds reduce the active surface area which is needed for the separation between P3 and P4.

Therefore the strategy for a chromatographic separation should be:

1) Remove everything from the crude mixture which does not need to be isolated by such a high-efficiency technique as chromatography.

2) Protect the main separation column from strongly binding compounds by using a pre-column or a pre-purification step.

3) Increase the separation efficiency between P3 and P4 by using different chromatographic elution modes, taking into account for example, counter-current or cross-current modes.

5.3.1
Crude Mixture Pre-Treatment

Preparative chromatography is a highly efficient technique but with a moderate time–space yield. Only those compounds, which really need the high efficiency for their isolation, should be subjected to the chromatographic column. Therefore

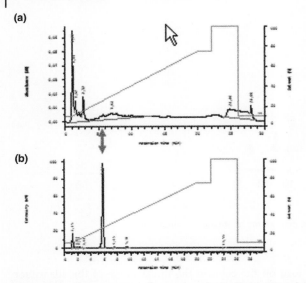

Figure 5.14 Chromatogram of natural product with (a) UV and (b) ELSD detection.

pre-treatment of the crude extract is essential. Figure 5.14 shows a chromatogram of a natural product.

Figure 5.15 gives an example of a pre-purified natural extract. The upper chromatogram shows the crude feed mixture with some late-eluting impurities. In these analytical chromatograms a gradient elution is used, which should be avoided in preparative chromatography because it requires higher effort in solvent management (solvent mixing, evaporation and re-adjustment). If the elution conditions are chosen properly the compound of interest elutes almost completely from the column while the impurities are strongly adsorbed. For this type of chromatography rather simple systems can be used consisting of low-pressure, open-bed columns packed with coarse silica particles (e.g., 63–200 µm). Most of the chromatographic systems are for single use because it is very difficult to re-adjust the bare silica for a second use. On the one hand the silica adsorbs water during chromatography and on the other hand the impurities are strongly adsorbed and thus also difficult to be removed from the silica surface. As the price for coarse normal phase silica is affordable, single use is often preferred over intensive solvent use for elution and re-equilibration. The lower chromatogram shows the crude mixture after it was subjected to a normal phase silica column on which the late-eluting compounds strongly adsorb.

It is important to obtain the correct ratio of crude feed mixture per amount of sorbent. In an example the crude mixture (Figure 5.15) was submitted to a silica gel column. The main compound should be isolated after the pre-treatment step with a purity of 80%. Increasing amounts of crude mass were injected onto the silica column. Fractions were taken and analyzed and the mass of the fractions with sufficient purity combined. It can be seen that a drastic drop in yield is

(a)

Original feed mixture

(b)

After filtration over silica all polar impurities are removed

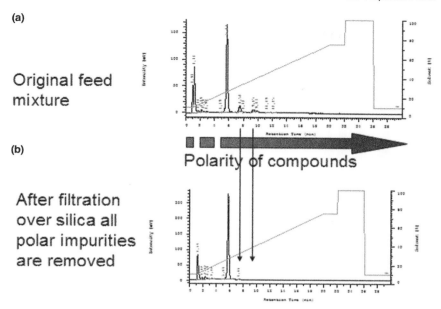

Figure 5.15 Chromatogram of (a) original feed and (b) after pre-purification.

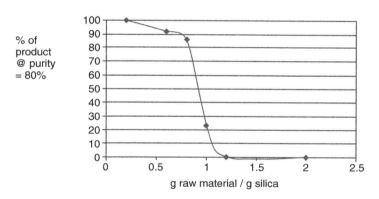

Figure 5.16 Ratio of silica mass per mass of product for pre-purification.

observed if the amount of crude submitted onto the silica exceeds a ratio of 1 g crude per g silica (Figure 5.16).

5.3.2
Final Purification of the End Product

After the late-eluting impurities have been removed by pre-treatment steps the mixture is reduced to a separation task to isolate the main compound (P3) shown in Figure 5.17.

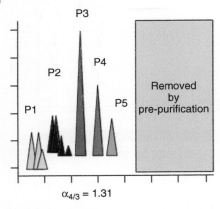

Figure 5.17 Reduced chromatogram by pre-purification.

Four main principles for high efficiency chromatographic separation should be considered:

5.3.2.1 Batch Chromatography

In single column batch chromatography portions of the feed mixture are injected into a packed bed of a certain adsorbent. During the elution with fresh eluent the compounds in the mixture are adsorbed differently on the adsorbent so that at the end of the column fractions of the pure compounds can be withdrawn from the system. While this approach is quite simple and straightforward it has some drawbacks. To obtain a high productivity in batch chromatography the amount of crude feed to be injected should be as high as possible. By increasing the load on a column the linear range of the adsorption isotherm, where there is no influence on the retention time or the peak shape of a single compound, is left and the chromatography is operated in the non-linear range of the adsorption isotherm. In this non-linear range there is a multitude of interferences between the different compounds of the mixture. These interferences can be measured and simulated but it is obvious that this is only possible for a limited number of compounds. As soon as the number of main compounds is higher than four or five the system gets much too complex. In practice this often leads to empirically, but not fully, optimized chromatographic systems where the amount injected is too low, resulting in low time–space yields, which means low productivities and high production costs. A good chromatographic purification system should thus be reduced to the compounds which need the high efficiency of modern preparative-scale chromatography. This system should then be operated at the highest load possible using stringent method development and rigorous system modeling and optimization [14, 15, 41, 42].

5.3.2.2 Annular Chromatography

One special case of chromatographic systems is the annular chromatography. In this type of column the adsorbent is packed into an annulus formed by two cylinders with different diameters, rotating around a vertical axis. While the introduction of the crude feed mixture is at one fixed point of the annulus, the whole radius of the annulus is flushed with fresh eluent. Due to the rotation of the packed bed zones of the single compounds develop in circular bands along the annulus. At the end of the adsorbent bed the compounds can be withdrawn in pure form at a certain outlet fraction. The system was first used in preparative paper chromatography in the 1950s. The most advanced use has been shown by Subramanian[43], who used the system for the continuous capture of a therapeutic protein. The biggest advantage of the system is given by the fact that it is continuously operated and can handle multiple fractions. Currently no system is available on the market from a commercial vendor. A special case of the annular chromatography is the use of several single columns mounted on a rotating valve and forming a column circle. Large systems of this type were developed by the US-company AST in the 1970s and operated with different examples, for example, for large-scale sugar or amino acid purification [2, 41, 42].

5.3.2.3 Steady-State Recycling Chromatography

A problem often faced, especially in natural product isolation, is the lack in resolution at the column outlet of a single column. When in batch chromatography two peaks still overlap at the end of the column, there are different possibilities to increase the resolution, but all of them (using smaller adsorbent particles or a longer column) result in a reduced pressure drop of the system. Higher pressure drop on the large scale is always related to higher equipment cost. In order to circumvent this problem the column outlet can be connected with the pump at the column inlet. With this set-up it is possible to re-direct the incompletely separated part of the chromatogram back to the column inlet where it can be separated in a second or third passage through the column. Band broadening due to diffusion is one practical problem with this approach, which can be overcome by peak-shaving. Peak-shaving is a technique where fractions are withdrawn from the system at the front and the rear part of the peak. Another problem in recycling chromatography is the fact that when peak-shaving is applied the amount of product in the column gets lower over time. While there is no need for fresh eluent during recycling there is nonetheless low productivity in this mode. To overcome this problem Miller *et al.* introduced a system where fresh feed is constantly injected into the non-separated part of the recycling chromatogram [2].

Figure 5.18 shows a recycling chromatogram for our separation problem. When the feed is introduced into the system P1 and P2 can be eluted from the column in normal batch elution mode. After P3 and P4 have eluted with insufficient resolution the system is switched to recycling mode and the fraction C1 is recycled. At the outlet of C2 a part of P4 is fractionated from the column (peak shaving), the rest is recycled to C3 where the procedure is repeated. At the end of fraction C4 much more P3 is still in the system because P4 has been withdrawn by the peak shaving technique. Both compounds are completely separated after C4 and P3 can now be withdrawn in pure form.

5.3.2.4 Simulated Moving Bed (SMB) Chromatography

The completely continuous and counter-current operation of a chromatographic separation is achieved with the simulated moving bed (SMB) approach [2, 14, 15]. This technology simulates the counter-current movement of stationary and mobile phase by shifting the inlet and outlet ports of a segmented chromatographic bed after pre-defined time intervals in the direction of the mobile phase flow [2, 14, 15, 41, 42, 43]. While SMB chromatography is mainly known in sugar separations (fructose/glucose-separation of high fructose corn syrup) and enantiomeric separations, some examples are also shown in the literature for natural product isolation like cyclosporine [2, 15]. If a separation problem can be reduced to a two-fraction separation, SMB chromatography can be applied in a straightforward manner. In our example a separation of the target compound P3 in the raffinate stream and P4 and P5 together in the extract stream can be achieved without modification of the classical SMB-setup, visualized in Figure 5.19.

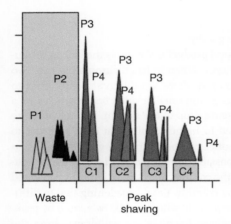

Figure 5.18 Chromatogram of recycling chromatography.

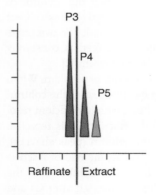

Figure 5.19 Chromatogram of simulated moving bed fractionation.

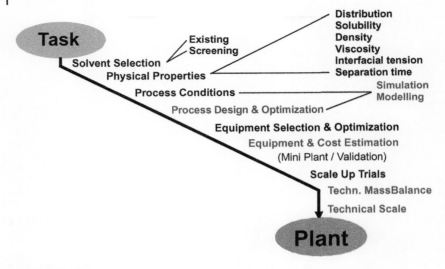

Figure 5.22 Work-packages from process development to detailed engineering.

thermal, mechanical influences etc. are unusual for chemical products, while they are regular with biological products. In the same way, biological mixtures have very similar behavior concerning their thermodynamic properties, which therefore requires quite complex separation processes.

Based on this, process development follows a defined operational sequence until the definition of the procedure and apparatus concept leads to conversion to the technical scale, shown in Figure 5.22. In that figure the processing steps "Process Design" and "Cost Estimation" constitute decisive points that define technical and economic continuation and adjustment as an process result. Preferably a broad parameter consideration is always aimed at arranging these results as exactly as possible.

It is evident that an extraction process always consists of concurrence of thermodynamic conditions and instrumental possibilities, which must be coordinated appropriately. The consideration of these details primarily appears in distribution coefficients, separation factors and in specific values of process equipment of the applicable extraction units. A summary is given in Figure 5.23.

5.4.2
Choice of Extracting Agent

The choice of an adequate extracting agent determines the effectiveness of all extracting processes. Matched to the mission, two basic objectives must be considered.

1) The application of extraction as a cleaning step removes annoying secondary components to achieve the quality requirements.

Selection of Extraction Agents (Solvents):

- Selectivity
- Capacity
- Reactive Agents

- Reaction Equilibrium
- Distribution Equilibrium

- Stability
- Solubility
- Capability of being regenerated
- Availability
- Toxicity
- Biotechnological compatibility

Figure 5.24 Checklist for parameters influencing selection of extraction agents.

application, solvents of the following groups can be used, where this listing gives examples geared to polarity and consequently it can account for the separation effort:

- ketones
- esters
- ethers
- alcohols
- aqueous two-phase systems.

With the exception of aqueous two-phase systems it concerns pure solvents with specific solution behavior, though maintaining biphasic properties or miscibility gap respectively. At the other hand, the aqueous two phase systems fundamentally consist of mutual soluble components (e.g., higher alcohols), and when adding salts it is forced into a biphasic state. Additionally, employing an adequate type of salt and concentration affects the required selectivity (s. also Section 5.3).

Concerning separation of phenolic compounds from an aqueous product stream, the dependence of the distribution from the solvent class is shown in Figure 5.25. The higher the gradient of the distribution curve, the higher is the affinity and capacity of the solvent and the higher is the performance.

5.4.3
Data of Chemical and Physical Properties

For conceptual design of a process, the following physical properties are needed:

- mixing and de-mixing time and degree
- distribution coefficients
- separation factor

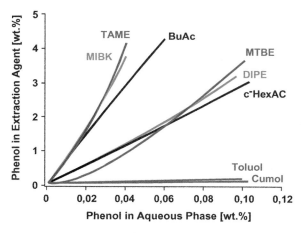

Figure 5.25 Dependence of extraction of phenolic compounds from water-like phase by different extraction solvents (MIBK methyl isobutyl ketone, TAME t-amyl methyl ether, BuAc butyl acetate, MTBE methyl t-butyl ether, DIPE di-isopropyl ether, cyclohexane acetate cHexAc, toluene, cumol).

- miscibility
- surface tension
- density
- mass transfer resistance (coefficient)
- eventually reaction stoichiometry and kinetics.

They are expressed as a function of concentration, by-product concentration and pH and/or conductivity due to for example, salts. Experimental parameter determination methods are described in detail in Refs [44–51].

5.4.4
Specification of Separation Sequence

Separation processes, especially those in the biological sector, require separation of components from multicomponent mixtures, which are usually very complex systems. These components present very similar thermodynamic behavior resulting in small separation factors, which implies aggravating circumstances,

An example is the separation of a specific antibody from a fermentation broth. The distribution curve for the target components and the proximate secondary components (key components) reflect an objective separation sequence, (Figure 5.26). If separation between key and secondary components is satisfactory, separation for any other components is also achieved. Based on components chromatograms, the respective equilibrium curves are plotted. Gradient slope and degree of overlapping of different components provide information about the complexity of the separation process which must be applied [52].

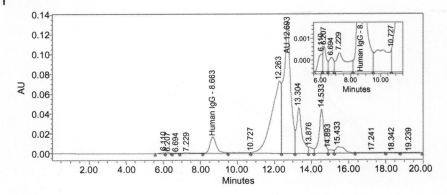

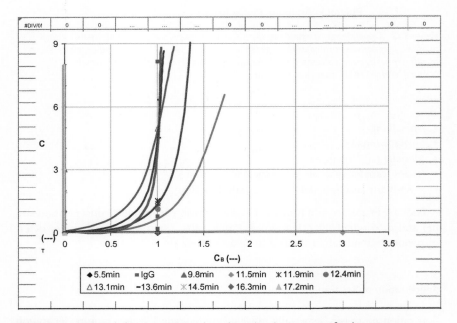

Figure 5.26 Analytical chromatogram and resulting distribution curve for the target components and the proximate secondary components (key components) for antibody separation from fermentation broth by ATP extraction.

5.4.5
Operation Concepts in Extraction

Having defined a separation sequence, procedural detail work to elaborate and define the process concept begins. For an extraction process the parameters and influencing variables are:

- distribution coefficient
- separation factor
- solvent charging
- target component concentration
- secondary components concentration (key component)
- proportion -main/secondary component.

Also, in the case of extractive treatment of plant extracts or juices, the common process alternatives form the basis for process design:

- single-stage discontinuous process control
- single-stage continuous process control
- multistage discontinuous cross flow extraction
- multistage continuous cross flow extraction
- multistage continuous counter flow extraction.

Typical concepts for extraction processes contain, in addition to the extraction step, a regeneration step of the extraction agent in order to recycle it and guarantee its re-usability. Regeneration of the extraction agent can be done with different unit operations for example, distillation, back-extraction, membranes etc. (Figure 5.27). Mass transfer of the target component(s) is directed from the feed phase by the extraction agent into the extraction phase. Thereby, distribution coefficient and separation factor are the dominating process design parameters.

Based on that, such process concepts can only be optimized by one objective, either yield or purity. Any optimization of yield must be gained by a quality

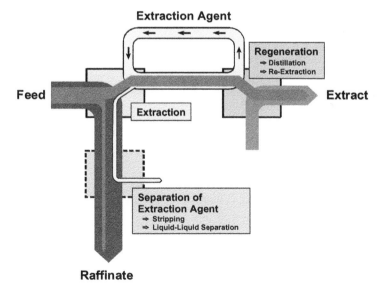

Figure 5.27 Extraction process concept.

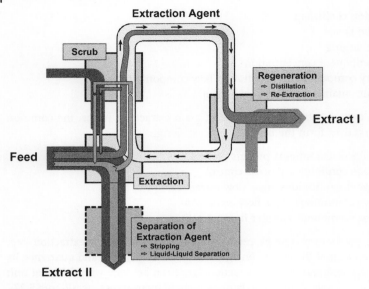

Figure 5.28 Fractional extraction process concept.

reduction and, vice versa, any quality improvement is necessarily accompanied by a yield reduction. To aspire to a result optimized in both yield and quality, the chosen process design must be based on the principle of fractional processing.

In contrast to the normal process concept before, the extraction step is now enlarged with an additional stage, the scrub stage, shown in Figure 5.28. Considering a fractional process, the difference in distribution coefficients (i.e., separation factor) in the extracting part of the process is used to transfer the target component preferentially into the extracting agent. In a post-washing stage, parts of secondary components, that are co-extracted due to separation factor, are "washed back". To avoid deficits caused during the purification of target component, the wash solution is finally led back to the initial mixture (feed).

When balancing the single phases, the thermodynamic advantage of this type of interconnection is evident. When the amount of the extracting solvent is normally lower than the amount of initial (feed) solution, the amount of wash solution is also lower than that of extracting agent.

Relating this conclusion to the mixture of a given feed and wash solution, the relevant process parameters vary in the following way, if the solute content in the feed rises:

- amount of extracting agent must be properly adjusted
- target component concentration declines
- fraction of secondary components rises
- there is a shift of the concentration ratio between target and secondary components.

The last-named effect is the most important one concerning the performance of the fractionating effect.

If a process concept is proposed independently of thermodynamic constraints, a more or less effective separation of target components will be achieved.

Related to the process scheme one component, with the higher affinity to water, remains in the water phase and the other target component will be drawn off due to its higher affinity to organics by the extraction phase as extract. Distribution coefficients and separation factors determine the internal recycling amounts and thereby equipment efforts necessary.

5.4.6
Determination of Equipment Efforts

The distribution coefficients of a system directly determine the necessary theoretical stages and, depending on the extraction amount, the equipment efforts required in order to achieve the prescribed yield criteria. To design an extraction unit and to choose the appropriate equipment type a decisive parameter is the mass transfer kinetics, which determines contact or residence time. It is dependent on:

- distribution coefficient
- diffusion velocity
- droplet size
- droplet building velocity
- mass transfer direction (mass transfer into the droplet or out of it)
- concentration difference
- mass transfer area.

A relevant determination of mass transfer kinetics with adapted parameter studies is only possible in single droplet experiments. In counter-current flow measurements cells all parameters could be determined independently [44–46]. Measurements at single droplets are advantageous against mixing cells or shaking experiments, since residence times can be controlled exactly, and are not affected by drop interaction and phase separation effects. In such measurement cells contact times of few seconds to some hours are possible. In addition, in such cells a limited number of droplets can simulate droplet swarm behavior. By this, the typical conditions of a real mixture can be simulated.

The core part of the single droplet measurement cell is a cuvette, through which the continuous phase flows. Against this flow direction the dispersed phase is injected by a nozzle as droplets which ascend until the droplet reaches equilibrium in velocity. The droplet stops at this position until, by interrupting the counter-current of the continuous phase flow, the droplet is drawn off at the outlet of the cuvette through a cone in a thin capillary, so that its composition can be analyzed offline. The number of droplets to be measured is given by the analytical method and therefore the necessary sample amount. Depending on the parameter adjusted at the measurement device, the mass transfer rate could be calculated as a basis for the determination of the mass transfer

Figure 5.29 Single droplet measurement unit (Bayer Technology Services GmbH, Leverkusen).

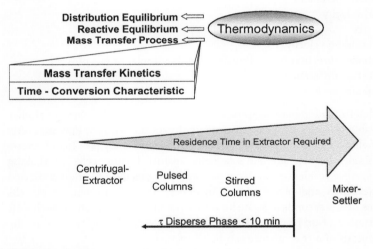

Figure 5.30 Residence time of LLE equipment types.

coefficients as a function of residence time. A detailed description is in Refs [44–51] (Figure 5.29).

Knowledge of mass transfer kinetics is one of the most important criteria for choosing the type of extraction equipment, based on necessary residence time, shown in Figure 5.30. On the one hand, with very low residence time, are the centrifugal type supported extractors and on the other hand, with very long residence times, are the volume supported mixing vessel type equipment like mixing-settler combinations. In between, with residence times of about 10 to 15 minutes

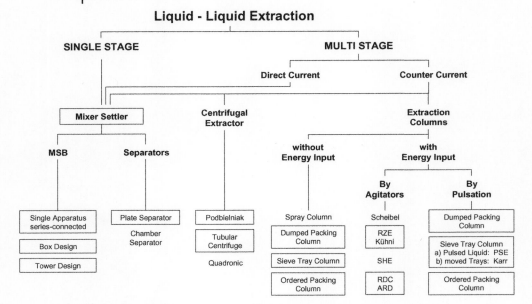

Figure 5.32 Decision tree for LLE equipment choice [53].

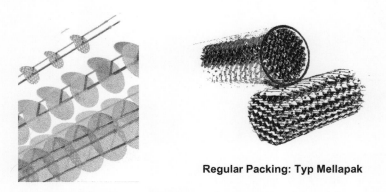

Figure 5.33 Sieve plate and packing internals for LLE columns.

It is known that the efficiency of any extraction columns is strongly affected by its specific load and difference to the maximal load, that is, the flooding point. Here, packed columns have an advantage as well. The operation region is much larger compared with the sieve plate column. In contrast to sieve plate columns, packed columns could be operated at low load values of about 20% to 25% without significant losses in efficiency degree. In the operation of sieve plate columns this loss of efficiency must be compensated by additional column length or additional extraction agent usage. In general it is complicated to realize low load values of lower than 40% in sieve plate columns.

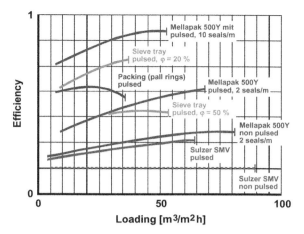

Figure 5.34 Efficiency of LLE columns.

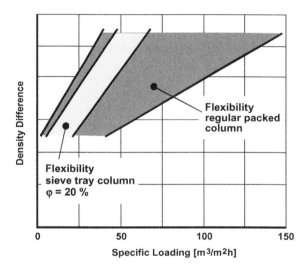

Figure 5.35 Operation range of sieve plate and packed columns.

Resumee: Packed columns have a broaoder operation load area and can be adapted more reliably and economically to the constraints of production plant.

5.5
Mini-Plant Technology for Extraction Process Development

In general mini-plant technology has the objective of substituting or at least reducing expensive pilot-plant experiments. Pilot-plant runs supply test amounts and/

or are the smallest equipment for each unit operation for a reliable scale-up into the necessary production scale [2, 54–60]. At product development, laboratory scale equipment is used which is in accordance with guidelines like pharmacopeia, nutrition or cosmetics guidelines or at least good state-of-the-art laboratory practice [61–63, 64].[1)-4)] Small first test material amounts must be generated. Methods must be reproducible, fast and easy to handle. No focus is given to the scale-up ability and cost efficiency.

During process development most small laboratory scale equipment is made of glass. Here, the focus is on scale-up ability and processing costs. In addition, not only are the single unit operations applied but the total process is operated in continuous mode to quantify enrichments, purge streams and long term stability. The appropriate equipment type of each unit operation must be chosen.

The task may differ if in addition test amounts for the customer market are to be supplied. Then, not the smallest scale-up able equipment for each unit operation is chosen which could be installed due to the resulting flow rates within a linked total process. Here, the boundary between mini-plant and pilot-plant becomes a smooth transition [2, 54].

Due to these demands, typical flow rates are about 1 to $2\,L\,h^{-1}$ at typical temperatures of about 5–80 °C and operation pressures of smaller 1 or 2 bar. Therefore, glass equipment is a standard. Based on that, typical dimensions suitable for scale-up are about i.d. 25–80 mm. An i.d. 50 is necessary to scale-up distillation equipment. To combine these flow rates with extraction equipment here an i.d. 30 is appropriate. Membrane, adsorption and chromatography equipment as well as precipitation/crystallization vessels are available at these corresponding flow rates. It should be taken into account that flow rates may differ by about a factor of 10 easily as a consequence of the position of the unit operation in the downstream schema. Normally the flow rates should be reduced for each following step. A proper temperature consistency within the whole mini-plant including fractionation and analysis is not easily achieved but a necessary precondition for reliable results.

Therefore, appropriate instrumentation and process control must be equipped. Samples should be taken at each step in all phases as a function of time in order to close the mass balance. At least analysis for dry solid mass, water content, target component, solvent concentration, for product specification relevant by-products and main plant contents like sugars, fats, proteins, essential oils and/or volatiles must be established within +/− 5% accuracy [32].

Exact flow and temperature control is important to gain reliable results for process design and model validation, therefore, double jackets must be constructed as consistent over equipment length as far as is technically possible. In addition, pH-control at each stage or inlet/outlet stream must be applied, if necessary in

1) www.ilumina-chemie.de (accessed 16 October 2009).
2) Schwobel, J. www.chemieplanet.de (accessed 16 October 2009).
3) Seilnacht, T. www.experimente.net (accessed 16 October 2009).
4) Seilnacht, T. www.seilnacht.com (accessed 16 October 2009).

Figure 5.36 Mixer-settler battery with 10 stages laboratory scale i.d. 50 mm (Bayer Technology Services GmbH).

dissociation or reactive extraction processes [27, 28, 49–51]. The following Figures 5.36–5.38 show laboratory scale equipment used for mini-plant technology applications.

Any configuration in the number of stages and any process concept with extraction, wash/scrub, re-extraction can be configured and operated. The stage efficiency degree of a mixer-settler is near to 1. Therefore, any stage prediction based on modeling which has been based on physical properties measured o the laboratory scale can be easily validated by taking samples between each stage in both phases. Valid data is only achieved by closing the mass balance for each component, which causes measurement of concentrations and volume and mass of both phases. In addition a stationary operation point is necessary, therefore, exact mass flow and hold-up volume control is a precondition.

To operate a mini-plant for process development and scale-up as well as test sample amount production a configuration is needed which consists of:

- Solid extraction equipment for maceration and percolation with about 50–1000 mL h^{-1}.
- LLE mixer-settler battery and 2–3 extraction columns for extraction/scrub/re-extraction in about 30–50 mm i.d. to be operated with and without pulsation and different internals like sieve plates, structured packing, rotary discs.
- At least 2 rectification columns i.d. 30–50 mm adiabatic design, to be equipped with different internals.
- Precipitation/crystallization vessel about 500–2.000 mL.

(a)

(b)

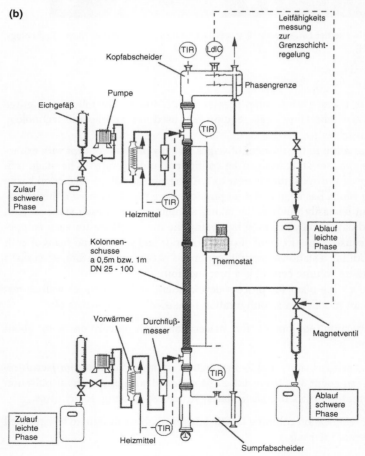

Figure 5.37 (a) Photograph and (b) P&I of sieve plate laboratory scale column i.d. 50 mm (Bayer Technology Services GmbH).

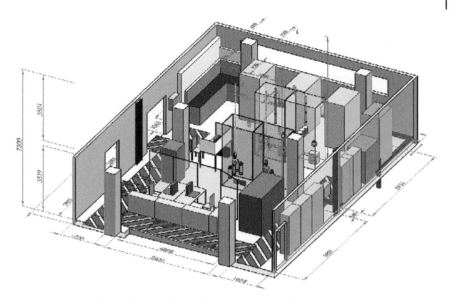

Figure 5.38 Plant layout for mini-plant technology (Institute for Separation and Process Technology, Clausthal University of Technology).

- Adsorption/chromatography units with about 10–200 mL min^{-1} flow rates, 10–100 bar and i.d. 10–75 mm.
- Solvent or buffer handling and mixing devices with about 10–15 vessels of about 100 to 1000 L, nitrogen overlaid and explosion proof facilities.
- Solvent recycling units like rotary evaporators about 25–50 L drums.

Each unit operation can be operated alone for model parameter determination and the unit operations can be operated in any sequence proposed by process conceptual design modeling. An example is given in Figure 5.38 which takes into account all relevant criteria of the bullet point list above, based on the state-of-the-art for further research activities within this area.

5.6
Cost Estimation

Cost estimation for determination of specific separation costs predicted based on process design results consists of:

- investment cost estimation CAPEX (capital cost expenses) by equipment cost multiplied with Lang-Factor 3–5 (CAPEX) [24] and
- operation cost calculation OPEX (operation cost expenses)

which results in both

- specific separation or production costs (COG) cost of goods in €kg^{-1} product in needed specification.

Typical ranges of costs per unit operation can be gained if proper process design and phase selection is fulfilled:

1) **Solid extraction**
   Solvent extraction [13, 65]: 10–100 €kg^{-1} at ton range and CO_2 [66, 67] may reach the same range.
2) **Liquid–liquid extraction**
   Typical specific separation costs for LLE s range in 0.1–10 €kg^{-1} [2].
3) **Chromatography**
   1–10 €kg^{-1} at proper phase screening and process design is possible [2].

Any individual exception—positive and negative—is of course possible. Individual calculations for each project are state-of-the-art [2, 24].

5.7
Total Process Development

Chemical engineering has developed systematic approaches for process development [2, 4, 7, 68–73]. Common is the combination of process simulation and experiments. Experiments to determine model parameters for each unit operation are done on the laboratory scale. Afterwards, conceptual process design is done by total process simulations and optimizations [22, 25–28, 46]. Mini-plant technology is applied to verify the proposed best process sequence experimentally. The main objective is to reduce piloting costs and determine recycling and purge strategies for production scale a priori.

5.7.1
Mini-Plant Technology

Mini-plant technology is described in general and detail in Refs [2, 54, 55, 60]. The transfer of these approaches which are more dedicated to classical chemical bulk industries to complex mixtures like plant-based extracts and biotechnology is not easy, but can be done [74].

As a consequence, to scale-down unit operations, which are later on used for a sound scale-up into production scale, typical operation conditions for the mini-plant system are:

- 1–2 L h^{-1} which results in about i.d. 30–50 mm equipment pieces;
- <300 °C;
- <10 bar, eventually 20 mbar;

- glass equipment to see phase behavior;
- solvent and auxiliary recycling applied and enrichments and purge strategies over time in continuous operation proven;
- cleaning procedures applied;
- each unit should be operated separately and in any combination sequence;
- external solvent storage, supply and recycling must be implemented;
- for proper model parameters determination total process measurement of all pressures, temperatures and mass flows as well as component concentrations in each stream and inside a few measurement points over column height (e.g., extraction and distillation);
- for piloting equipment at one scale higher must be available.

Standard mini-plant for distillation and steam distillation is described for example, in Refs [32–34, 74–87].[5)–13)] The critical step in plant-based extraction process development is SLE combined with pre-treatment steps prior and extract purification steps after extraction. Easily 60–90% of total costs are linked to these steps – if the raw material is not expensive [3].

For SLE mini-plants a typical Soxhlet with specifications is shown in Figure 5.39. A Soxhlet like this is not feasible for model parameter determination because the reflux rate and therefore residence time are not controlled and have to be measured. It is only useful to produce first product samples. Additionally two limiting

Laboratory Scale

- **Capacity:**
  - 10 – 40 g Raw Material
  - 100 – 200 ml Solvent
- **Diameter: 50 mm**
- **Height: 180 mm**
- **Temperature up to 100 °C**
- **Retention Time up to 6 hours**

Figure 5.39 Soxhlet mini-plant equipment [10].

5) www.glassapparate.de (accessed 16 October 2009).
6) www.qvf.de (accessed 16 October 2009).
7) www.dedietrich.com (accessed 16 October 2009).
8) www.kuehni.ch (accessed 16 October 2009).
9) www.buchi.com (accessed 16 October 2009).
10) www.rotationsverdampfer.com (accessed 16 October 2009).
11) www.artisansoftwaretools.com (accessed 16 October 2009).
12) www.wikipedia.de (accessed 16 October 2009).
13) www.gea.com (accessed 16 October 2009).

Figure 5.40 Maceration mini-plant equipment [10].

Laboratory Scale

- **Capacity:**
  - 1 – 20 g RawMaterial
  - 50 – 100 ml Solvent
- **Diameter: 120 mm**
- **Height: 120 mm**
- **Retention Time up to several days**

Figure 5.41 Percolation mini-plant equipment [10, 13].

Laboratory Scale

- **Capacity:**
  - 30 – 50 g RawMaterial
  - 0,01 – 10 ml/min Solvent
- **Inner-Diameter: 25 mm**
- **Height: 200 mm**
- **Temperatureup to 100 °C**

cases exist, which is either totally mixed, like maceration, or totally flow-controlled, like percolation. Typical laboratory equipment and their specifications are shown in Figures 5.40 and 5.41.

Finally they can be combined into a standard equipment for SLE parameter determination, which is described in Figure 5.42. Either maceration or percolation can be used with this equipment. Soxhlet is applied by combining percolation for dedicated residence times and fluxes with specific evaporation and condensation devices separately.

Online detection by diode array detection (DAD) and refractive index (RI) or evaporating light-scattering detection (ELSD) is fruitful. Afterwards, fractions are collected for off-line structure and amount analysis by GC-MS, HPLC-MS, GC-FID and NMR. In addition, sample amounts can be generated for 2D-HPLC fractionation which leads to pure component fractions for either screening on efficacy or structure analysis or both.

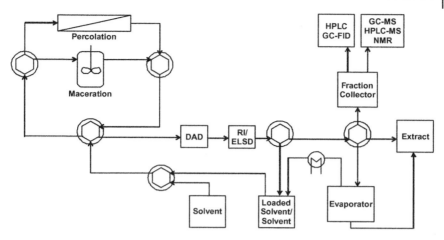

Figure 5.42 Standard laboratory equipment [10, 13].

Solvent recycling is done on the laboratory scale mainly by rotation evaporators in 1 to 25 L vessels [2][14)–16)] and in pilot or production scale more often by falling film or thin film evaporators.[17)–19)]

In general the main tasks to gain valid data for process design are (i) to run repetitive experiments to quantify reproducibility and (ii) to close the mass balance for each component group, like target component/group, by-product/group, volatile compounds, water, solvent, dry raw material etc. The mass balance is more accurate if both phases are analyzed compared to the usual case of balancing the second phase upon initial values. This, in combination with proper calibration methods, requires most laboratory input using classical methods like:

- Karl-Fischer titration for water content in liquid samples
- GC/HPLC for component analysis
- toluene extraction for water content determination in solids
- drying: residual moisture and dry solid weight as reference basis
- GC for solvent content of liquid samples
- hydro-distillation for content of volatile organic compounds (VOC's) in raw materials
- evaporation (<5–10 mbar) for residual content determination of samples from 1 to 100 mL

14) www.kuehni.ch (accessed 16 October 2009).
15) www.buchi.com (accessed 16 October 2009).
16) www.rotationsverdampfer.com (accessed 16 October 2009).
17) www.artisansoftwaretools.com (accessed 16 October 2009).
18) www.gea.com (accessed 16 October 2009).
19) www.loemi.com (accessed 16 October 2009).

- centrifugation and filtration for solid content determination of liquid fractions 1–100 mL
- evaporation (at 10 mbar or below) and condensation with cooling and into closed sample vessels/vials to prevent unspecific solvent evaporation over experimental time [10, 13, 23].

5.7.2
Examples of Typical Processes

5.7.2.1 Quassia Extracts

Wood of the neotropical species *Quassia amara* was used as test system for experimental investigations. The object of the experiments was the selective desorption of two quassinoid compounds: quassin and neoquassin. Quassinoids are terpernoid compounds. Chemically, these molecules are seco-triterpene-δ-lactones. They have important pharmaceutical and insecticidal properties and their intensely bitter nature make them ideal as bittering agents for beverage and foodstuffs.

Wood samples were prepared by milling and sieving the supplied dried coarse chips to obtain fractions of different mean particle sizes. These fractions were stored below room temperature. The desorption solvent was a 20% (w/w) water/ethanol solution. Besides the target components, by-products are also desorbed. The desorbed solution was therefore regarded as a three components mixture of quassin (TC1), neoquassin (TC2) and by-products combined under SC. All chemicals were obtained from Merck (Darmstadt, Germany) at analytical grade.

Percolation experiments were carried out in a fixed bed with a diameter of 2.5 cm and a height of 36.5 cm corresponding to a volume of 179 mL. The bed was filled with 50 g wood chips (particle size 0.25 mm) and extracted for one hour. The ratio of solvent mass used to wood mass was approximately 12. To compare the obtained results, maceration experiments were performed with the same extraction time and mass ratio of solvent to wood. For the raw material in use, no serious differences were observed between maceration and percolation results in the laboratory experiments. For the chosen conditions, equitable TC1 yields are obtained for both operational modes. Slightly higher TC2 and total yields are observed when extracting under maceration conditions.

In the downstream process, a purification of the extracted solution was carried out by chromatographic separation and LLE. Although the investigations were not very detailed, they allow the potential of these unit operations for the purification of the extract solution to be assessed. Before chromatographic separation, the extract solution obtained after desorption was first concentrated. A sample of the concentrated extract solution was then separated analytically in a HPLC column. The obtained fractions were then analyzed. Before chromatographic analysis, by-products in the concentrated extract solution constituted 92% (w/w) of the extracted mass. The fractions obtained after chromatographic separation are nearly pure solutions, containing only TC1 and TC2. The fraction containing TC1 has a purity of 93%. Chromatography can thus crucially contribute to

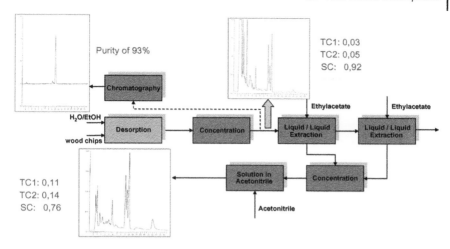

Figure 5.43 Process scheme for *Quassia* purification by chromatography as a comparison w/o pre-purification by LLE [12].

the extract solution purification. Chromatography is scaled up from the analytical method to preparative and technical scale by increasing the particle diameter of the adsorbents to 15 μm and maximizing the feed amount injected and the flow rate.

For LLE experiments, the concentrated extract solution was twice successively extracted with ethyl acetate in the cross-current operational mode. In the first stage an ethyl acetate (w)/extract solution (w) ratio of 3:1 was used. The aqueous phase obtained after this stage was then re-extracted with ethyl acetate using a ratio of 1:1. The organic phases of both stages were mixed and concentrated. Since ethyl acetate presence in the solution to be analyzed damages the HPLC packing material, the obtained extract was dissolved in acetonitrile and analyzed. The by-products concentration amounts 76% (w/w) after LLE. Compared with concentrations before extraction, the TC1 and TC2 concentration in the extract obtained after LLE are increased by 213%. So, coupling of desorption, extraction and chromatography in this order of separation sequence is the efficient combination for the production of TC1 and TC2 [12] (Figure 5.43).

5.7.2.2 Spruce Bark Extracts

The use of substances of natural origin in nutritional, pharmaceutical and cosmetic products becomes more and more important. The sources of these products are leaves, blooms, branches, bark, roots, seeds and fruits. The content of the value component lies in the range of 0.3 to 3% with up- and downturns due to the year and the growing area [88]. Further variations occur because of the crushing, the transport and the digestion of the raw material. This has a serious impact, especially for the wood and the fruits; therefore mainly the leaves are used for the extraction.

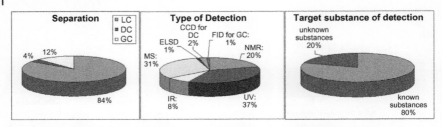

Figure 5.44 Distribution of literature articles concerning separation and substance analysis.

On the way to the final product the critical steps digestion, SLE and purification must be passed through. The digestion step and its quality has a high effect on the economy of the process. In this part the target components must be made accessible, which implies a gentle procedure due to the low thermal stability of most of the components. This gentle mode of operation continues over the whole process.

For the example of the spruce bark a method for the extraction and identification of value components from natural material was developed. The aim of this work was to get an overview which techniques are suitable to solve this given problem, which boundary conditions must be considered and how long the developed techniques and standard operational procedures need. To get an overview of the subject's analytical separation and substance identification a literature research was made in which over 5200 abstracts were found. Of 195 more precisely evaluated abstracts 112 were relevant to the topic. An overview regarding the separation and the substance identification is shown by the following distribution in Figure 5.44.

In the chosen example, spruce bark, the concentrated extract was separated by 2D liquid chromatography. A variation of the solvent and the gradient causes different selectivity in the two dimensions of the chromatographic separation sequence. Hence it was possible to separate even those substances which eluted at the same time in the first dimension of the chromatographic separation. The application of mathematical methods to substance characterization with UV-spectra databases shows that the comparability of UV-spectra is constrained by the operating conditions of the HPLC. Comparability is only possible if the operation conditions like pH, pressure, temperature etc. are exactly the same. Therefore a database for the substance identification of already known substances should be based on a more sensitive technique such as mass spectroscopy (MS) or NMR. Also the application of a UV-spectra based mathematical tool for the determination of the purity of a peak fails too due to the fact, that in plant extracts many substances with very similar UV-spectra are present. It was possible to identify several substances with NMR and MS. The main substance could be identified as taxifolin-3'-O-glucoside. Furthermore, several flavonoid–glucoside derivates of stilbene–glucoside dimers were found (Figure 5.45).

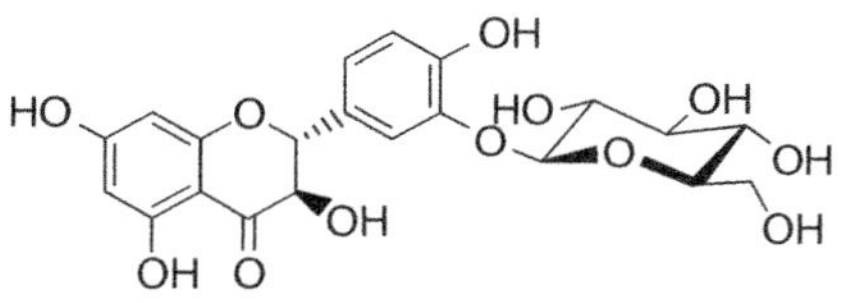

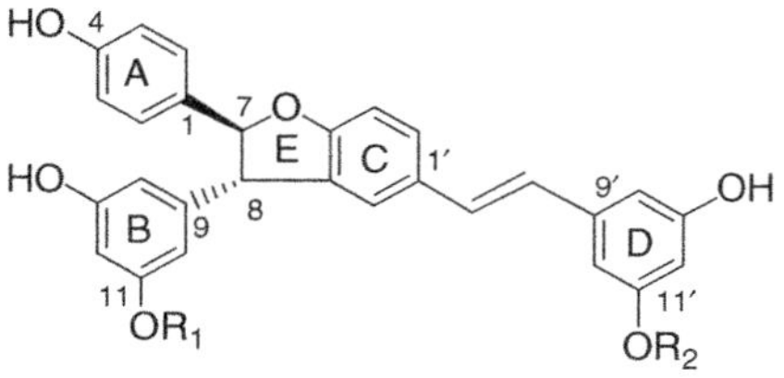

Figure 5.45 Stilbenes.

5.7.2.3 Pepper Extracts

Pepper is the most widely traded spice in the world with an annual production of about 270 000 tons. Botanically it can be differentiated between chilli peppers and peppercorns, which share the pepper market in approximately equal parts by weight. The world's largest producers are Vietnam, India, Brazil, Indonesia, Malaysia, Sri Lanka, China and Thailand. The main use of pepper is for food seasoning, whereas only a small amount is used for extraction.

Pepper extraction is carried out with supercritical fluids like carbon dioxide to produce essential oils as well as with solvents like ethanol or ethyl acetate to produce for example, piperine and capsaicin. While supercritical fluid extraction results in essential oil, the extract obtained by solvent extraction needs downstream processing. After the SLE step the extract is concentrated by solvent evaporation under vacuum to save temperature sensitive compounds or LLE. Finally purification is carried out with thin-film evaporators, fractionated distillation or crystallization [89].

5.7.2.4 Vanilla Extracts

Out of more than 100 species of the tropical climbing orchid of plant genus *Vanilla* today only three are of commercial interest: *Vanilla planifolia* Andrews (*Vanilla fragrans* Salisbury), *Vanilla thaitiensis* JW Moore and *Vanilla pompona* Schiede. Because of its pod quality and yield of the main aromatic constituent vanillin *V. planifolia* is the most cultivated [90]. It is widely assumed that the Aztecs were the first to use vanilla as medicine and spice. In the early eighteenth century the Spanish brought vanilla to Europe where cultivation in greenhouses was possible but due to absence of natural pollinators the plants neither blossomed nor produced fruit [91, 92]. In 1941 artificial pollination by hand was discovered and successfully established in Reunion Island. From there vanilla cuttings were taken to the Indian Ocean Region and onwards worldwide [93]. Nowadays, Madagascar,

The Comoros and Reunion Island grow 80% of the world production. Indonesia, Mexico, Uganda, India, Papua New Guinea and China are further main exporters and Europe, the USA and Japan are the main importers. The world annual production of cured Vanilla is estimated to be 2000–3000 tons [92]. The market share of natural vanilla is about 5%. The lion's share of the annual demand of vanillin is chemically synthesized from guaiacol in many industrialized countries.

Green vanilla pods do not have the typical aroma of vanilla and contain only little or no vanillin. The characteristic aroma is obtained through a curing process during which vanillin and other aroma constituents are developed as a result of different biochemical transformations. During the process of curing the typical flavor is obtained by hydrolysis of nonvolatile glycosidic aroma precursors mainly composed of glucovanillin by a β-glucosidase [93]. The process of curing is done in several different ways depending on the region, though all curing methods involve four basic steps: wilting or killing, sweating, drying and conditioning [91, 93].

In the killing process mature pods are put in about 60 °C heated water for three minutes which ends the respiratory metabolism. After killing, the pods begin to turn a typical brown color. Immediately after killing the sweating phase begins during which the pods are stored in cloth-lined wooden boxes for about one day at ambient temperature. A rapid dehydration as well as a slow fermentation will start during which the enzymatic reactions occur. Afterwards the pods are entirely of brown color.

The step of drying lasts for three months during which pods come to final moisture of 20–25%. The pods are sun-dried during the mornings and stored in the shade in the afternoons. Conditioning of the dry pods takes place for some months in closed boxes. After this terminal process pods are full of their characteristic fragrance and ready for market [91, 93]. Cured vanilla pods are shipped from exporting countries whole or cut. Further processing takes place in the importing countries. In Europe vanilla is mostly sold in whole beans while in the USA vanilla is primarily processed to ethanolic solvent extracts.

5.7.2.5 Taxol Extracts

In 1955 the Cancer Chemotherapy National Service Center (CCNSC) which was founded by the National Cancer Institute (NSI) started its work. The function of this institute was to execute analysis to determine the potential against cancer of a huge number of different substances. At first time only substances produced by fermentation were tested; but after 50 years secondary plant metabolites were included. In the early 60 years this screening showed that the bark of the pacific yew tree (*Taxus brevifolia*) has a cytotoxic potential. In 1967 Wani et al. isolated the relevant property substance and its molecular structure. This was published in 1971 [94–96].

A sufficient supply of the active ingredient was not possible for a long time, because there were not enough pacific yew tree. Furthermore it was necessary to extract ca. 1200 kg bark to get only 10 g of this substance. Nevertheless in the last 70 years the discovery of its effectiveness against leukemia and its property to

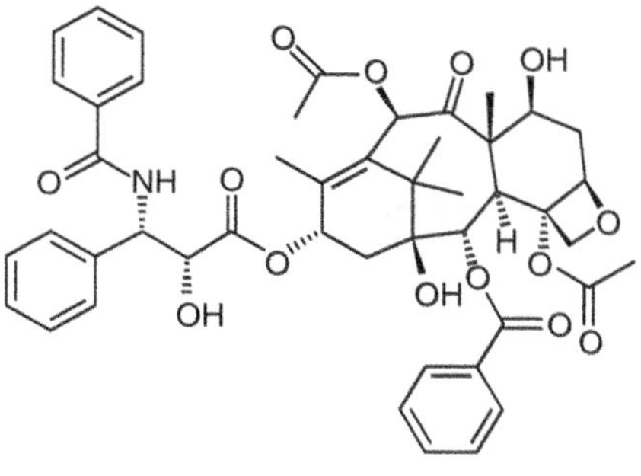

Figure 5.46 Paclitaxel.

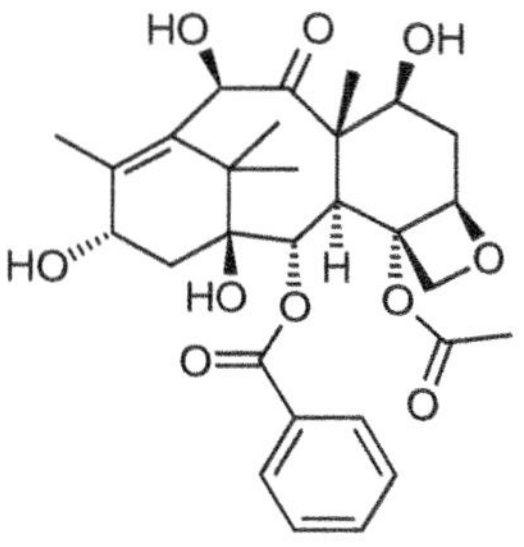

Figure 5.47 10-Deacetylbaccatin.

stabilize microtubules accelerated the further development and the commercial launch of this product. In 1992 Bristol–Meyers–Squibb received the FDA-admittance for paclitaxel (Figure 5.46) as a drug against metastasized ovarian carcinoma.

In 1988 Pierre Potier published a method for the semisynthetic production of paclitaxel. This method started with the very complex molecule 10-deacetylbaccatin (Figure 5.47) which is present in high concentrations (up to 0.1 mass-%) in the needles of the European yew tree. This substance can be isolated by a SLE with ethanol and a following LLE with dichloromethane and water [97, 98]. In 1989 Robert Holton developed a process with a highly increased yield and 1992 he patented a technique with an overall yield of 80%. On the basis of this patent it was possible to produce paclitaxel in a sufficient amount to economically justify the cost [99] (for the synthetic route see Figure 5.48). The patent rights were lodged by Bristol–Meyer–Squibb. State-of-the-art production process is that SLE is followed by LLE and final purification with SMB chromatography [100].[20]

Typical ingredients of yew tree are:

- diterpene esters (N-containing, as well as N-free)
- bi-flavonoids
- lignan derivatives

20) Novasep www.novasep.com (accessed 16 October 2009).

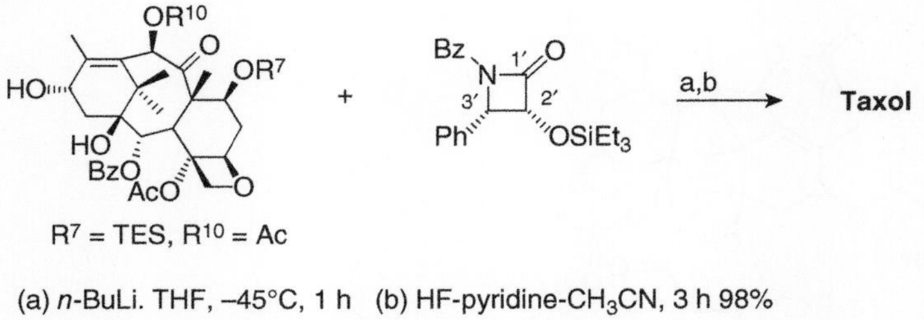

Figure 5.48 Production of paclitaxel by Bristol–Meyer–Squibb [99].

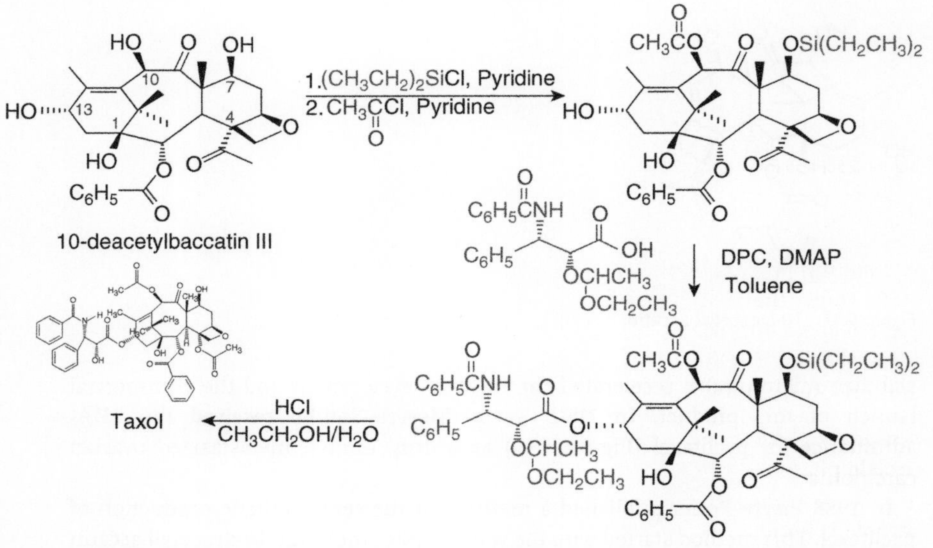

(a) n-BuLi. THF, −45°C, 1 h   (b) HF-pyridine-CH$_3$CN, 3 h 98%

Figure 5.49 Semisynthetic Taxol production from 10-deacetylbaccatin-III [101, 102].

- cyanogen derivatives
- sequoite
- phytoecdysteroids

and *Taxus baccata* leaves which consist of a mixture of different ester alkaloids (analysis of alkaloids only in water-based extracts is complicated, because they are not well-soluble in water).

The semisynthetic production route is summarized in Figure 5.49. *T. baccata* contain typically amounts of 0.4–1.7% (needles), 0.9% in seed, 0.08–0.1% in the bark with following derivatives:

- ester-alkaloids (Taxol™ A/ Taxol B, as well as 11 other Taxol derivates)
- N-free Taxol derivatives

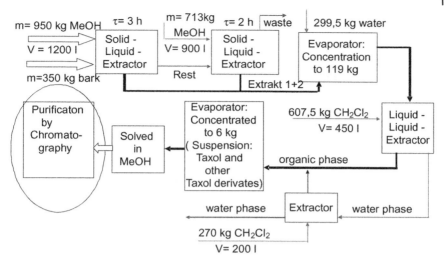

Figure 5.50 Taxol production flowsheet.

A resulting process flowsheet for Taxol manufacturing is summarized in Figure 5.50. There, the feed is based on a bark extract (and not needles) which gives a yield of about 40 g~12%. This is lower than the semisynthetic route via baccatin-III from needles. The final purification by RP chromatography after solid extraction and pre-purification via LLE is done using the following conditions:

stationary phase: LiChrosphere RP18 (5 μm)
mobile phase: 0–54 min. MeOH/H$_2$O
54–75 min. MeOH
sample volume: 10 μL MeOH
flow rate: 1.0 mL min^{-1}.
detection: 220 nm

Applying this method produces the results summarized in Figure 5.51. This is a typical example which is called up into production due to its economic benefits by SMB chromatography [100]. Detailed literature on that topic can be found in Refs [102–105].

5.7.3
Process Modeling and Scale-Up

Search of scientific literature about SLE modeling resulted in almost 36 000 hits which may be reduced finally to 100 sensible articles. The articles dealing with modeling of plant-based extraction processes can be divided into three domains: about 20% of the articles apply design of experiments to develop statistical models; 70% of the articles use distributed plug flow models assuming porous particles or

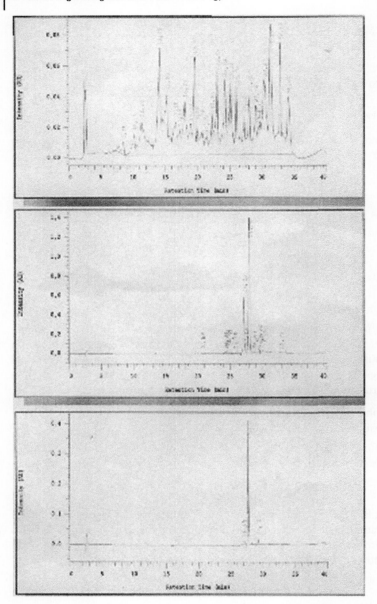

Figure 5.51 Chromatograms of raw feed, after LLE and final product after chromatography.

shrinking core models; and only 10% of all reviewed articles integrate the botanical aspects of the raw material for their model.

The rigorous models for plant-based extraction can be divided into two types: models dealing with SLE only on the macro-scale, that is, the equipment level, and models also taking the micro-scale, on the single particle or plant cell level, into account. The equipment level is characterized by geometric dimensions and process parameters. Usually, the macro-scale can be described by either (i) a model of a distributed plug flow column with or without axial dispersion:

$$\frac{\partial c_L}{\partial t} = D_{ax}\frac{\partial^2 c_L}{\partial x^2} - v\frac{\partial c_L}{\partial x} - k_{eff}a_p(c_L - c_S) \qquad (5.1)$$

or a continuous stirred tank reactor:

$$\frac{\partial c_L}{\partial t} = -k_{eff}a_p(c_L - c_S) \qquad (5.2)$$

where

- c_L concentration of component i in liquid phase
- c_S concentration of component i at solid phase surface
- t time coordinate
- x axial space coordinate
- D_{ax} axial dispersion coefficient
- v velocity axial dispersion
- k_{eff} efficient mass transfer coefficient (linear driving force)
- a_p particle surface for mass transfer

In special cases like hydro-distillation other models, like a fluidized bed, must be chosen to simulate plant particles together with boiling water and steam bubbles.

The rigorous plant-based extraction models for micro-scale can basically be divided into three different approaches: (i) shrinking core model; (ii) model of broken and intact cells; and (iii) model of desorption equilibrium:

(i) The shrinking core model tries to describe the micro-scale kinetics according to a model used in chemical engineering for gas-solid reactions. The particles are assumed to be spherical. Desorbing first the oil in the outer zone of the particle is followed by the solute of the inner core diffusing to the outer zone until it is transported out of the particle. Therefore the desorption front in the spherical particles moves from the surface to the core, which appears like a shrinking core. The desorption process is finished when the shrinking core's volume reaches zero.

$$\frac{\partial q}{\partial t} = \frac{3 \cdot k_f}{r}(c_L - c_S(r)) \qquad (5.3)$$

where

- q load of component i in solid phase
- $c_S(r)$ concentration of component i at solid phase surface
- k_f film mass transfer coefficient.

(ii) Due to mechanical pre-treatment, like milling, some of the plant cells containing the oil are crushed, which leads to the model of broken and intact cells. It is assumed that the oil from the broken cells (free oil) is directly accessible for the solvent whereas the oil from the intact cells (tied oil) has to pass cell walls, which leads to different mass transfer resistances. Therefore extraction curves can be divided into three sections:

- Constant extraction rate: particles are covered with free oil and main mass transfer resistance is in the solvent phase.
- Failure extraction rate: the free oil is already desorbed at the extractor entrance and diffusion mechanism of bounded oil starts.
- Free oil is totally desorbed and the process is limited by intra-particle diffusion [106].

(iii) The third approach applying adsorption/desorption equilibria relates to the modeling of chromatography processes. The particles are assumed to have pores in which the target compound is adsorbed. For the SLE process the solvent has to diffuse through the external film and the pore, extract the component and find the way back out of the particle into the bulk fluid. In the pores the adsorption/desorption equilibrium can be described by any isotherm for example, Henry, BET, Langmuir, Freundlich. Besides that, the component can be homogenously distributed in the particle or it can be represented by some other type of distribution for example, a parabolic profile [2]:

$$c_S(x,r,t) = b_1(x,t) + b_2(x,t) \cdot r^2 \tag{5.4}$$

where

$b_1(x, t)$ first coefficient for parabolic concentration profile of component i in solid particle

$b_2(x, t)$ second coefficient for parabolic concentration profile of component i in solid particle

The maturity of modeling of solid extraction has not yet reached to same degree as other unit operations like LLE or chromatography. Therefore, further research is needed to integrate this unit into total process conceptual design and optimization studies on an adequate level [10, 13].

Figure 5.52 points out (see arrows) that with typical moderate target component contents of about 0.01–0.1% products with about 10–100 € kg^{-1} product separation cost of goods are economically achieved at yearly production amounts of a few tons if solid extraction, with additional LLE and final purification by chromatography units can be operated under pre-purified and therefore easier and more economic conditions with fewer by-products. Investment and operation costs are calculated based on the process design modeling. With these separation cost scenarios, cost prediction of products for typical agrochemical, nutrition, cosmetic, or even high cost pharmaceuticals applications can be generated.

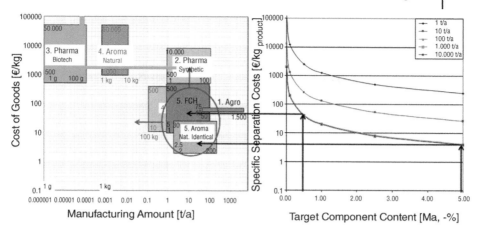

Figure 5.52 Cost of goods for compound extraction of spruce bark by SLE, LLE and chromatography based on total process modeling combined with laboratory scale experiments for model parameter determination [2, 10, 13].

5.8
Future Developments

In future process development especially the first proposal of a process conceptual design will be more efficient and faster if a sound method for feed characterization could be established. This is coincident with the availability of reliable analytical methods based on small mg amounts, like GC-MS, HPLC-MS and NMR. The existing mini-plant technology must be further equipped with more process analytical measurement technologies—hopefully online DAD, RI or ELSD, pH and conductivity measurements—as well as automation by PCS. In addition to that, thermodynamic theory has the challenge to deliver physical properties, since actually with 97% of the components, data are missing [22].

Acknowledgments

Jochen Strube would like to thank his former colleagues, J.-P. Josch, E. Ndocko and S. Sommerfeld, at Bayer Technology Services GmbH/Leverkusen and M. Kassing and F. Grote at Clausthal University of Technology as well as to PD Dr. R. Hänsch, TU Braunschweig for laser scanning microscopy and Prof. Dr. H. Schulz, Julius-Kühn-Institute, Quedlinburg for Raman mapping as cooperation partners giving botanical background measurements and fruitful explanations. Also thankfully acknowledged is Prof. Dr. B. Schneider, University of Jena for HPLC-MS and Prof. Dr. D. Kaufmann, Clausthal University of Technology, for NMR and GC-MS measurements.

References

1 Harborne, J.B. (1999) *Phytochemical Dictionary*, 2nd edn, Taylor & Francis Ltd, London, UK.
2 Goedecke, R. (Ed.) (2006) *Fluid-Verfahrenstechnik*, Bd. 1 u. 2, 1st edn, Wiley-VCH Verlag GmbH, Weinheim, Germany.
3 Sattler, K. (2007) *Thermische Trennverfahren*, 1st edn, Wiley-VCH Verlag GmbH, Weinheim, Germany.
4 Perry, R.H. and Green, D.W. (1997) *Perry's Chemical Engineers Handbook*, 7th edn, McGraw Hill, Columbus, Ohio, USA.
5 Schönbucher, A. (2002) *Thermische Verfahrenstechnik*, 1st edn, Springer, Berlin, Germany.
6 Mersmann, A., Kind, M. and Stichlmair, J. (2005) *Thermische Verfahrenstechnik*, 2nd edn, Springer Verlag, Berlin, Germany.
7 Bohnet, M. (ed.) (2003) *Ullmann's Encyclopedia of Industrial Chemistry*, 6th edn, Wiley-VCH Verlag GmbH, Weinheim, Germany.
8 Lohrengel, B. (2007) *Einführung in die thermischen Trennverfahren*, Oldenburg Wissenschaftsverlag, München, Germany.
9 Jenelten, U., et al. (2009) Phytoextraction from a flavors and fragrances industry perspective, Symposium with Workshop, 2009-10-30 Dechema e.V., Frankfurt at working party Phytoextracts.
10 Kaßing, M., Jenelten, U., Schenk, J. and Strube, J. (2010) *Chem. Eng. Tech.*, 33 (3), 377.
11 Kaßing, M., Grote, F. and Strube, J. (2008) *Chem. Ing. Tech.*, 80 (9), 1031.
12 Ndocko, E., Bäcker, W. and Strube, J. (2008) *Sep. Sci. Technol.*, 43, 642.
13 Kaßing, M., Strube, J., Jenelten, U. and Schenk, J. (2009) *Chem. Ing. Tech.*, 81 (8), 1064.
14 Unger, K.K. and Weber, E. (1994) *Handbuch der HPLC I + II*, 1st edn, GIT Verlag, Darmstadt, Germany.
15 Strube, J. (2000) *Technische Chromatographie*, Shaker Verlag, Aachen, Germany.
16 Gmehling, J. and Kolbe, B. (1992) *Thermodynamik*, 2nd edn, Wiley-VCH Verlag GmbH, Weinheim, Germany.
17 Hansen, C.M. and Skaarup, K. (1967) *J. Paint Technol.*, **39**, 511.
18 Hildebrand, J.H. and Scott, R.S. (1951) *J. Phys. Chem.*, **55** (4), 619.
19 Maurer, G. (ed.) (2004) *Thermodynamic Properties of Complex Fluid Mixtures*, Wiley-VCH Verlag GmbH, Weinheim, Germany.
20 Gross, J. and Sadowski, G. (2001) *Ind. Eng. Chem. Res.*, **40**, 1244.
21 Ondruschka, B. and Klemm, W. (2008) *Chem. Ing. Tech.*, **80** (5), 803.
22 Josch, J.P., Bäcker, W. and Strube, J. (2009) *Chem. Ing. Tech.*, **81** (8), 1065.
23 Strube, J. (2010) Destillation, in *Handbuch des Arznei- und Gewürzpflanzenanbaus II* (ed. B. Hoppe), Saluplanta e.V., Bernburg, Germany.
24 Peters, M.S., Timmerhaus, K.D. and West, R.E. (2003) *Plant Design and Economics for Chemical Engineers*, 5th edn, McGraw-Hill, Columbus, Ohio, USA.
25 Ströhlein, G., Schulte, M. and Strube, J. (2003) *Sep. Sci. Tech.*, **34** (14), 3353.
26 Strube, J., Gärtner, R. and Schulte, M. (2002) *Chem. Eng. J.*, **85**, 273.
27 Franke, M., Gorak, A. and Strube, J. (2004) *Chem. Ing. Tech.*, **76** (3), 199.
28 Franke, M.B., et al. (2008) *AIChE J*, **54** (11), 2925.
29 Wehlau, A. (2010) *Lebensmittel- und Futtermittelgesetzbuch (LFGB)* Carl Heymanns, Verlag, Berlin, Germany.
30 Bund für Lebensmittelrecht (1991) *Technische Hilfsstoffverordnung-THV*, BGBI, Berlin, Germany.
31 Pflanzenschutzgesetz (PflSchG), Berlin (2009) http://www.jusline.de/Pflanzenschutzgesetz_28PflSchG29.html (accessed 16 October 2009).
32 Hoppe, B. (ed.) (2010) *Handbuch des Arznei- und Gewürzpflanzenanbaus II*, 1st edn, Saluplanta e.V., Bernburg, Germany.
33 *Deutschen Arzneibuch (DAB)* (2008) Bundesgesundheitsministerium,

Bundesinstitut für Arzneimittel und Medizinprodukte BfArM, Berlin, Germany.

34 *Europäischen Arzneibuch* (2007) (6th edn with revision 6.3), Bundesgesundheitsministerium, Bundesinstitut für Arzneimittel und Medizinprodukte BfArM, Berlin, Germany.

35 *Homöopathischen Arzneibuch (HAB)* (2008) Bundesgesundheitsministerium, Bundesinstitut für Arzneimittel und Medizinprodukte BfArM, Berlin, Germany.

36 Anfärbereagentien für DC und PC, Merck KGaA, Darmstadt, www.merck.com (accessed 16 October 2009).

37 Van Berkel, G.J., Ford, M.J. and Deibel, M.A. (2005) *Anal. Chem.*, **77**, 1207.

38 Van Berkel, G.J. and Kertesz, V. (2006) *Anal. Chem.*, **78**, 4938.

39 Pabst, G. (ed.) (1999) *Köhler's Atlas der Medizinal-Pflanzen*, Bechtermünz Verlag, Eltville, Germany.

40 Heinrich, M. (2006) *Walnuss dietary Inhaltsstoffe: Pharmacological Research 2005, 52, Understanding local Mediterranean diets: a multidisciplinary pharmacological and ethnobotanical approach*, S. Karger AG, Basel.

41 Subramanian, G. (ed.) (2007) *Bioseparation and Bioprocessing*, 2nd edn, Wiley-VCH Verlag GmbH, Weinheim, Germany.

42 Vogel, J., et al (2002) *Biotechnol. Bioeng.*, **80** (5), 559.

43 Subramanian, G. (ed.) (2006) *Chiral Separation Techniques*, 3rd edn, Wiley-VCH Verlag GmbH, Weinheim, Germany.

44 Schröter, J., Bäcker, W. and Hampe, M. (1998) *Chem. Ing. Tech.*, **70** (3), 279.

45 Henschke, M. (1995) *Dimensionierung liegender Flüssig-flüssig-Abscheider anhand diskontinuierlicher Absetzversuche*, VDI-Verlag, Düsseldorf, Germany.

46 Henschke, M. (2004) *Auslegung pulsierter Siebboden-Extraktionskolonnen*, Shaker-Verlag, Aachen, Germany.

47 Bäcker, W., Schäfer, J.-P. and Schröter, J. (1991) *Chem. Ing. Tech.*, **63** (10), 1008.

48 Reissinger, K.-H., Schröter, J. and Bäcker, W. (1981) *Chem. Ing. Tech.*, **53** (8), 607.

49 Leistner, J. (2004) Modellierung und Modellparameterbestimmung von Extraktionsprozessen. Dissertation, TU Dortmund.

50 Manski, R., et al (2006) *Chem. Eng. Tech.*, **29** (12), 1513.

51 Franke, M. (2006) *Optimierung hybrider Verfahren*, VDI-Verlag, Düsseldorf, Germany.

52 Rosa, P.A.J., et al. (2009) *J. Chrom. A*, **1216** (50), 8741.

53 Brandt, H.W., Reissinger, K.-H. and Schröter, J. (1978) *Chem. Ing. Tech.*, **50** (5), 345.

54 Deibele, L. and Dohrn, R. (2006) *Miniplant-Technik*, 1st edn, Wiley-VCH Verlag GmbH, Weinheim, Germany.

55 Steude, H.E., Deibele, L. and Schröter, J. (1997) *Chem. Ing. Tech.*, **69** (5), 623.

56 Schwister, K. (2007) *Taschenbuch der Verfahrenstechnik*, 3rd edn, Carl Hanser Verlag, München, Germany.

57 Ignatowitz, E. and Fastert, G. (2009) *Chemietechnik*, 9th edn, Verlag Europa-Lehrmittel, Haan-Gruiten, Germany.

58 Dialer, K., Leschonski, K. and Onken, U. (1986) *Grundzüge der Verfahrenstechnik und Reaktionstechnik*, Carl Hanser Verlag, München, Germany.

59 Vauck, W.R.A. and Müller, H.A. (2000) *Grundoperationen chemischer Verfahrenstechnik*, 11th edn, Wiley-VCH Verlag GmbH, Weinheim, Germany.

60 Hofen, W., Körfer, M. and Zetztmann, K. (1990) *Chem. Ing. Tech.*, **62** (10), 80.

61 Schwetlick, K. (2009) *Organikum*, 23rd edn, Wiley-VCH Verlag GmbH, Weinheim, Germany.

62 Lucas, M. (2000) Untersuchung der Ursachen von Aromaveränderungen an einem alkoholischen Heilkräuterdestillat während der Reifeperiode, Dissertation in der Fak. III der TU-Berlin, www.tu-berlin.de (accessed 16 October 2009).

63 Wich, P. (2001) Experimentalchemie, www.experimentalchemie.de (accessed 16 October 2009).

64 Voigt, R. (2000) *Pharmazeutische Technologie*, Deutscher Apotheker Verlag, Stuttgart, Germany.
65 Rinder, R. and Bomme, U. (1998) *Wasserdampf-Destillation ätherischer Öle aus frischen Pflanzen oder angewelkten Pflanzen*, Bayrische Landesanstalt für Landwirtschaft (LfL), Freising-Weihenstephan, Germany, 1st edn, www.LfL.bayern.de (accessed 16 October 2009).
66 Marckmann, H., et al. (2008) *Chem. Ing. Tech.*, **80** (9), 1306.
67 Lack, E. and Seidlitz, H. (2009) New industrial applications of supercritical fluid technology for extraction and generation of micro-powders. Paper at AChEMA '09 session Phytoextracts, 2009-05-15, Frankfurt.
68 Blaß, E. (1997) *Entwicklung verfahrenstechnischer Prozesse*, Springer Verlag, Berlin, Germany.
69 Simmrock, K.H. and Fried, B. (1990) *Chem. Ing. Tech.*, **62** (12), 1018.
70 Stephanopoulos, G. (1986) *Chemical Process Control*, Pearson, New Jersey, USA.
71 Biegler, L.T., Westerberg, A.W. and Grossmann, I.E. (1997) *Systematic Methods of Chemical Process Design*, Prentice Hall, New Jersey, USA.
72 Ng, K. (2002) *Chemical Process Equipment*, Butterworth-Heinemann, Oxford, UK.
73 Marquardt, W. (ed.) (2008) *Process Systems Engineering*, Elsevier, Amsterdam, The Netherlands.
74 Strube, J. (2008) *Mini- und Mikro-Plant Versuchsanlage der Thermischen Verfahrenstechnik*, Vortrag bei Dechema Fachausschuss Extraktion und Phytoextrakte, Clausthal-Zellerfeld, Germany.
75 Atkins, P.W. (2001) *Physikalische Chemie*, 3rd edn, Wiley-VCH Verlag GmbH, Weinheim, Germany.
76 Wedler, G. (2004) *Lehrbuch der Physikalischen Chemie*, 5th edn, Wiley-VCH Verlag GmbH, Weinheim, Germany.
77 Försterling, H.-D. and Kuhn, H. (1991) *Praxis der Physikalischen Chemie*, 3rd edn, Wiley-VCH Verlag GmbH, Weinheim, Germany.
78 Bergmann, L. and Schaefer, C. (2008) *Lehrbuch der Experimentalphysik*, Bd. 1, 12th edn, Springer Verlag, Berlin, Germany.
79 Winnacker, K., Küchler, L. and Dittmeyer, R. (2006) *Chemische Technik*, Band 1–8, 5th edn, Wiley-VCH Verlag GmbH, Weinheim, Germany.
80 Gmehling, J. and Brehm, A. (1996) *Grundoperationen aus M. Baerns (Hrsg.), Lehrbuch der technischen Chemie*, Bd. 2, 1st edn, Thieme Verlag, Stuttgart, Germany.
81 Bierwerth, W. (2007) *Tabellenbuch Chemietechnik*, 6th edn, Verlag Europa-Lehrmittel, Haan-Gruiten.
82 Seider, W.D., Seader, J.D. and Lewin, D.R. (1999) *Process Design Principles*, 1st edn, John Wiley & Sons, Inc., New York, USA.
83 Weiß, S. (Hrsg.), 1996.*Thermisches Trennen*, 2nd edn, Deutscher Verlag für Grundstoffindustrie, Stuttgart, Germany.
84 Baehr, H.D. and Stephan, K. (2008) *Wärme- und Stoffübertragung*, 6th edn, Springer, Berlin, Germany.
85 Bird, R.B., Stewart, W.E. and Lightfoot, E.N. (2007) *Transport Phenomena*, 2nd edn, John Wiley & Sons, Inc., New York, USA.
86 Kutscher, S. (1995) *Verfahrenstechnische Berechnungen*, 1st edn, VDI-Verlag, Düsseldorf, Germany.
87 Moore, W.J. and Hummel, D.O. (1986) *Physikalische Chemie*, 1st edn, Walter de Gruyter, Berlin, Germany.
88 Bäcker, W., et al. (2003) Konzeptpapier des Arbeitskreises Phytoextrakte, Dechema e.V, www.dechema.de (accessed 16 October 2009).
89 Karvy Comtrade Limited (2008) Seasonal Outlook Report Pepper (PDF)., www.karveycomtrade.com (accessed 16 October 2009).
90 Sinha, A.K., Sharma, U.K. and Sharma, N. (2008) *Int. J. Food Sci. Nutr.*, **59** (4), 299.
91 Francis, F.J. (1999) *Encyclopedia of Food Science and Technology – Vanilla*, Wiley-VCH Verlag GmbH, Weinheim, Germany.
92 Kahane, R., et al. (2008) *Chron. Horticult.*, **48** (2), 23.

93 Odoux, E., Chauwin, A. and Brillouet, J.-M. (2003) *J. Agric. Food Chem.*, **51**, 3168.
94 Goodman, J. and Walsh, V. (2001) *The Story of Taxol: Nature and Politics in the Pursuit of An Anti-Cancer Drug*, Cambridge University Press, Cambridge, UK.
95 Wani, M.C., Taylor, H.L., Wall, M.E., Coggon, P. and McPhail, A.T. (1971) *J. Am. Chem. Soc.*, **93** (9), 2325.
96 Nicolaou, K.C. and Guy, R.K. (1995) *Angew. Chem.*, **107**, 2247.
97 Gueritte-Voegelein, F., *et al.* (1986) *Tetrahedron*, **42** (16), 4451.
98 Hook, I., *et al.* (1999) *Phytochemistry*, **52**, 1041.
99 Holton, R.A. (Florida State University) (1992) Method for preparation of taxol using beta-lactam, US Patent Nr. US5,175,315.
100 Schulte, M., Britsch, L. and Strube, J. (2000) *Acta Biotechnol.*, **20** (1), 3.
101 Bartsch, V. (2000) *Das Taxol-Buch*, Thieme, Stuttgart, Germany.
102 Canell, R.J.P. (1998) *Natural Products Isolation*, Humana Press, Totowa, New Jersey, USA.
103 Blaschek, W., *et al.* (1992) *Hagers Handbuch der Naturstoffe*, 5. Auflage; Band 4-6. Springer Verlag.
104 Steglich, E., Fugmann, B. and Lang-Fugmann, S. (1996) *Roempp-Lexica Chemie*, Thieme, Stuttgart, Germany.
105 Steglich, E., Fugmann, B. and Lang-Fugmann, S. (2000) *Natural Products*, Thieme, Stuttgart, Germany.
106 Sovova, H. (2006) Modeling of supercritical fluid extraction of bioactives from plant materials, (Eng) in *Functional Foods Ingredients and Nutraceuticals* (ed. J. Shi), CRC Press, New York, USA, pp. 76–109.

6
Extraction Technology

Andreas Pfennig, Dirk Delinski, Wilhelm Johannisbauer and Horst Josten

6.1
Introduction

Solid–liquid extraction (SLE) of components from natural sources has been practiced since ancient times. Extraction by large-scale industrial equipment has been done since the early twentieth century. Major applications were and still are the production of lipids from oil seeds, sucrose from sugar beet and phytochemicals from plants. Additional applications have been made accessible recently like the production of aromas and spices, caffeine from coffee beans, or functional hydrocolloids from algae. Solid–liquid extraction is also applied for the production of metals from ores or the removal of undesirable impurities or toxins, but these are outside the scope of this book.

In this chapter we will describe the basics for extractor design. The different extraction principles will be presented as well as theoretical and experimental methods to characterize extraction performance as basis for extractor design. Insight into industrial equipment is given together with considerations relevant for the selection of the optimal extractor.

When designing equipment for extraction from solid natural products it should be kept in mind that the extraction step is incorporated into a more complex flowsheet, which usually contains solvent recycling as well as pre-treatment and downstream process steps (Figure 6.1). Parts of these typical process steps will also be dealt with in this chapter.

6.2
Extraction Process Basics

6.2.1
Introduction

The extraction of plant material with a suitable solvent or solvent mixture consists of a series of elementary steps. Here for brevity the term "solvent" is used to

Industrial Scale Natural Products Extraction, First Edition. Edited by Hans-Jörg Bart, Stephan Pilz.
© 2011 Wiley-VCH Verlag GmbH & Co. KGaA. Published 2011 by Wiley-VCH Verlag GmbH & Co. KGaA.

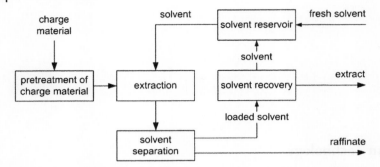

Figure 6.1 Integration of a solid–liquid extraction step into a typical process flow sheet.

include solvent mixtures. In many cases the raw material has been pre-treated for stabilization including drying to defined moisture content. When the liquid solvent is contacted with the solid plant material it intrudes into the solid, often supported by capillary forces. Since the solid is generally porous this may be a very fast step. The liquid can then dissolve components that are accessible within the capillaries as, for example, is the case for vanillin in vanilla [1]. In a next step the solvent intrudes into the cellular structure of the plant material, dissolving further fractions of components. Simultaneously the cells may swell, increasing the mass-transfer resistance in the capillaries. The soluble components are then dissolved by the solvent. Some of the components that are soluble in principle may be trapped inside the solid particles by strong adsorptive or chemisorptive forces. The dissolved components must then be transferred in the opposite direction from within the cell structures to the capillaries and then to the fluid surrounding the solid particle, where the convective mass transfer is usually significantly faster than the essentially diffusive processes inside the particle. An additional mass-transfer resistance may result from local equilibrium within the solid particle since the solubility limit may be exceeded. In some cases the plant material may have been compacted into pellets, which leads to corresponding additional mass-transfer steps on a larger scale.

The mass-transfer rate strongly depends on the details of the structure of the plant material. Seeds may for example, have a protecting peel that introduces an additional mass-transfer hindrance. To reduce mass-transfer resistance the size of the particles may be decreased and special structures may be broken by grinding. For the technical process the contact time between solvent and plant material must be sufficient to allow mass transfer to properly take place. This can be achieved by appropriate configuration of the equipment ensuring intense contact between the phases.

When designing the equipment it is helpful to have suitable models at hand describing the processes occurring. These should include the different size scales ranging from the particle to the equipment scale. The model parameters should be determined experimentally, especially because plant material shows high variability in structure and content of desired as well as undesired components for example, changing with the season of harvesting, location of production as well as conditions of storage and pre-treatment (see Chapter 1). Thus in the first part of this chapter the various basic aspects will be dealt with, including the different extraction principles, determination of target component and selection of solvents, as well as pre-treatment and storage of the material.

6.2.2
Extraction Techniques

6.2.2.1 Solid–Liquid Extraction

Common to all SLE methods is the intimate contact of solids and liquid. The following two different major principles can in general be distinguished [2]:

- percolation (from latin "percolare", to trickle through) and
- immersion (from latin "immergere", to dip in).

Additionally maceration can be differentiated from immersion and, mostly for the determination of solubilities on the laboratory scale, Soxhlet extraction can be performed.

Percolation For percolation the solid material is stacked as a fixed bed and the solvent passes through this bed, generally driven by gravity from top to bottom in technical equipment. To increase the overall contact time the solvent can be recycled and be passed several times through the fixed bed until sufficient mass transfer has taken place. The free volume of the plant material must be high enough to allow free passage of the solvent through the fixed bed. Advantages of this process are the relatively low mechanical stress on the solid material as well as the filtering capabilities of the fixed bed, ensuring that the extract is relatively free of solids content. If the extraction is performed with fresh solvent, leaching of the desired components can be essentially complete.

Immersion An alternative arrangement is to immerse the solid particles in the stirred solvent, which ensures very intense contact between the phases. Thus the solids are completely "dipped into" the liquid. In this case the mechanical stress on the particles is relatively high and the filtering of the extract must be achieved separately. At most an equilibrium between solid and solvent can be achieved, which, depending on the solubility properties, may leave a significant amount of the desired components remaining in the solid.

Both percolation and immersion have advantages and disadvantages, which will be discussed later. All available industrial equipment can be classified into these two contact types (see Section 6.5.4).

Maceration A third method is referred to as maceration, which means "soaking" of a solid in a liquid. Maceration and immersion are quite similar, because

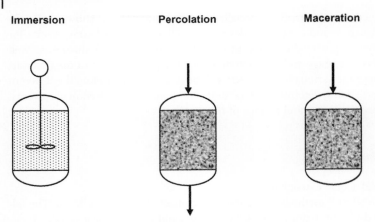

Figure 6.2 Basic principles for solid–liquid extraction.

with both methods the solids are completely dipped into the liquid. However in immersion the liquid and the solid may be in motion with each other, while in maceration the solid is just contacted with liquid without motion of liquid or solid for a defined residence time. These three basic principles for SLE are illustrated in Figure 6.2.

Soxhlet Extraction Soxhlet extraction is generally performed in the laboratory to determine the content of all components that can be dissolved in the solvent. Typical equipment is shown in Figure 6.3. The solid material to be extracted is placed in a reservoir of filter material inside chamber (5). The solvent in the bottom reservoir (1) is then heated to boiling and condensed at the condenser (9). The reflux from the condenser passes the extraction chamber where it comes into contact with the solid to be extracted at a temperature dependent on the boiling point of the solvent. After a certain amount of solvent has collected in this chamber the chamber is emptied through a special siphon arrangement (6) to the side of the chamber and flows back into the distillation flask. This process repeats automatically until the heating of the bottom ceases. By this arrangement the solid is contacted with fresh solvent until in principle all soluble components are extracted. The dissolved components collect and are concentrated in the distillation flask. Thus this equipment is very suitable to determine the content of soluble component in a sample of solid material. It should be kept in mind though that not all of the soluble components may be accessible under these conditions due to adsorptive or chemisorptive forces. Also degradation of the components may occur in the distillation flask, since this is kept at the boiling temperature of the solvent for considerable time. Thus in some instances more of the desired components

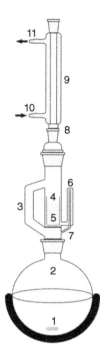

Figure 6.3 Diagram of a Soxhlet extractor. 1: Stirrer bar, 2: Still pot, 3: Distillation path, 4: Thimble, 5: Solid, 6: Siphon top, 7: Siphon exit, 8: Expansion adapter, 9: Condenser, 10: Cooling water in, 11: Cooling water out.

may be extracted at for example, high-temperature conditions in an immersion process.

6.2.2.2 Process Alternatives

To be able to optimize the mass-transfer and equilibrium behavior of the extraction system supercritical fluid extraction (SFE) can be used. Under supercritical or near-critical conditions of the solvent the density of the system can be varied continuously over a very wide range between liquid and gas conditions. Especially solubility, viscosity, and diffusivity – all of which are mainly determined by density – are key parameters determining the extraction behavior. While in an appropriately chosen liquid solvent solubility is generally sufficiently high, high viscosity and low diffusivity hinder a fast extraction process. A vapor phase would significantly reduce mass-transfer resistance, solubility would be minimal. Thus super- and near-critical conditions allow optimizing between these limiting cases. Often carbon dioxide is used for SFE due to its specific properties, to which modifiers can be added to adjust polarity. Since the critical point can be influenced by the dissolved components, a certain distance of the operating conditions to the critical point should be ensured. The details of SFE are discussed in Chapter 4 of this book.

The advantages of SFE can partly also be exploited when ordinary solvent extraction is pressurized. This leads to a situation where temperature can be increased so as to decrease mass-transfer resistance and optimize solubility of the solutes. Also the higher pressure facilitates the liquid penetrating the solids. The higher pressure is achieved by pressurizing the extractor, for example, with an inert gas like nitrogen, leading to so-called pressurized solvent extraction (pressurized fluid extraction, PFE).

Additionally mass transfer can be assisted and equilibrium can be influenced by supplying energy to the extractor. Two options that are technically applied are microwave-assisted extraction (MAE) and ultrasonic extraction (sonication).

6.2.2.3 Continuous or Discontinuous Extraction

Solid–liquid extraction can be performed in continuous as well as in discontinuous or batch mode. As in all separation processes counter-current continuous mode will lead to the highest concentration gradients and thus the most efficient use of the solvent. Due to the specific properties of the solid such a process is often not feasible. In SLE another operation mode is possible, where the solvent trickles through the solid material. The solvent can be directly run through the equipment several times so as to increase the amount of transfer component collected, which leads to a batch-like operation mode. Alternatively fresh solvent can be percolated through the solid until the desired limiting concentration is reached.

6.2.3
Further Considerations

To investigate SLE processes experimentally or theoretically for process design, several additional considerations should be taken into account. Extracting plant material generally will lead to a wide variety of components that will be extracted. Usually it requires quite considerable effort to set up a quantitative analytical method to measure the desired compositions. Often chromatographic techniques like HPLC are used. In some cases more than one analytical method is required [3]. Also the identification of components can be problematic. In the optimum case the desired component is identified and can be analyzed quantitatively. This requires a pure standard and a good analytical signal. This often cannot be achieved. Then several key or indicator components can be chosen alternatively. If the components are not identified, only relative concentration information can be gained. Nevertheless quantitative but relative insight into partition coefficients and extraction kinetics is possible.

In preparing the experiments care must be taken that the sample material to be extracted is stable. Due to the sensitivity of plant components to for example, oxidation this can be a significant challenge. The stability can be tested by analyzing the content of the desired component extracted from the same material at different times and by analyzing the identical sample on various days. This ensures that neither are the desired components degraded inside the sample material as time proceeds, nor do the components interact with the solvent or other ingredient of the analytical procedure. Stability can be increased by keeping the samples at low temperature, preferably under inert atmosphere. Care should also be taken to check the mass balances to detect possible problems.

When planning the experiments it should be remembered that the space of potential operation conditions often is highly multidimensional. Parameters include the nature of selected solvent, solvent composition including the concentration of modifiers (for example, also water content in alcohols), temperature, particle size, pre-treatment of the plant material. Considerations for an optimum choice of parameters will be presented in Section 6.5 on industrial extraction, but they should already be considered during the experimental investigations and modeling activities.

6.3
Experimental Procedures

6.3.1
Extraction Curves

In order to characterize extraction kinetics, extraction curves can be used that represent the concentration of the extracted component in the solvent versus extraction time. Here and in the following only one component is referred to for brevity, although usually more than one component is extracted simultaneously. In that case all relevant components will be considered independently or in their interdependence, especially if relative kinetics are to be used for separation via kinetic selectivity. For SFE it is also common to plot the cumulated amount of extract versus extraction time or amount of solvent passed through the extractor [4].

Since these extraction curves are the basis for conventional extractor design the experimental procedure for determination of this characterization of SLE kinetics will further be discussed.

6.3.2
Determination of Equilibrium and Kinetics

There are several methods to determine the equilibrium and kinetics of the extraction of natural raw materials. The easiest and most commonly applied method on laboratory scale is the experiment performed in a stirred vessel. A known amount of solid material is contacted with a defined amount of solvent in a beaker and stirred at defined conditions for example, with respect to stirring intensity and temperature. Possibly a temperature jacket for temperature control can be applied. The extraction kinetics can be determined by taking samples during the extraction process at defined time intervals. The samples are then analyzed with respect to the transfer component. If the samples taken are small compared with the overall amount of solvent in the beaker, no correction for the amount of withdrawn

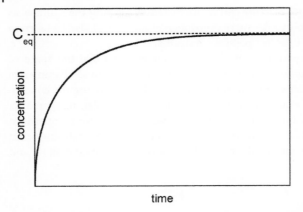

Figure 6.4 Schematic representation of the extraction curve for the stirred-vessel experiment corresponding to immersion.

sample is needed. The determined concentration in the samples can then be plotted versus extraction time.

A typical extraction curve obtained in this way is shown in Figure 6.4. The overall appearance is an exponential approach to an asymptotic limit. By comparing the experimental points to the results of model equations a deeper insight into the basic dependencies can be obtained. If for example, a certain amount of transfer component is readily available in the macro pores of the particles that readily dissolves, the extraction curve will have a quick, jump-like increase at the beginning and the curve depicted in Figure 6.4 will appear to start at a non-zero concentration. If on the other hand the particles have a wide particle-size distribution, the extraction curve will have a steeper slope at the beginning due to the small particles present and then level off significantly more slowly towards equilibrium, due to the large particles present in the sample, leading to the impression that equilibrium cannot properly be reached.

Equilibrium is reached when the concentration in the extract becomes independent of extraction time and the extraction curve levels off asymptotically reaching the upper limit. The dependence of this upper limit on solid-to-solvent ration is characteristic for different solubility behavior. If the solubility limit of the solute in the solvent is reached, it is independent of the amount of solvent added as shown in Figure 6.5. If the equilibrium concentration depends inversely on the amount of solvent, the solute very readily dissolves in the solvent. In that case the solvent is available in excess and the equilibrium curve in Figure 6.5 shows a slope of -1 in the log–log plot. Between these limiting regimes the concentration can in principle be described by a solid–liquid partition coefficient. In order to determine this equilibrium partition coefficient the balance should generally be set up to determine the amount of solute remaining in the solid and the amount of solvent that cannot be removed from the solid must be taken into account.

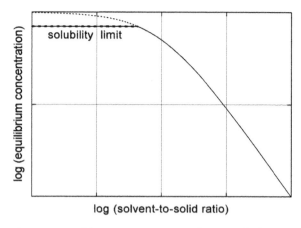

Figure 6.5 Equilibrium concentration as function of solvent-to-solid ratio.

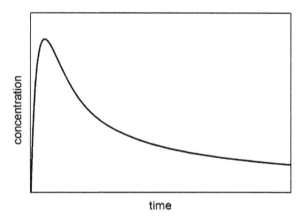

Figure 6.6 Schematic extraction curve corresponding to a percolation experiment.

This stirred-vessel approach described corresponds to immersion extraction. Alternatively a percolation method can be used where fresh solvent passes through a fixed bed. On the other hand also for the stirred vessel a fresh solvent stream can be added and the corresponding extract stream removed continuously. Investigation of these different operational modes reveals different details of the extraction process.

In the overall process to characterize the extraction kinetics obviously only the interplay between micro and macro kinetics can be observed. Here micro kinetics is characteristic for the kinetic processes directly around and inside the solid particles, while macro kinetics characterizes the influence of the macroscopic equipment on the extraction curve. For a laboratory-scale percolation experiment for example, an extraction curve as shown in Figure 6.6 is obtained. The overall concentration will generally be much lower in the case of percolation as compared with immersion.

6.3.3
Standard Laboratory Experiment

The extraction process is obviously determined by a variety of influencing system-specific parameters as well as operating conditions of the experiment. The system-specific parameters include the diffusion coefficient of the solute, the partition coefficient, some characterization of the porosity of the solids, etc. Simultaneously the macroscopic set-up of the experiment influences the extraction curve. It has been shown that performing experiments for different macroscopic configurations will allow discrimination of the different influences on the micro kinetics [5]. Such a comparison of the results of different macroscopic experiments also allows discrimination of the detailed models describing the micro kinetics.

Thus in order to be able to perform such experiments easily and systematically a standard laboratory equipment for SLE has been proposed, that allows realization of the different macro kinetics with minimal effort, and is shown in Figure 6.7. Within the temperature-controlled compartment a stirred vessel as well as a fixed-bed column are contained. Both can be connected either to a feed of fresh solvent or the solvent can be circulated through both cells. In either case the outlet stream is passed through an online analysis that can for example, be a UV-Vis spectrometer. Alternatively samples can be withdrawn at different times.

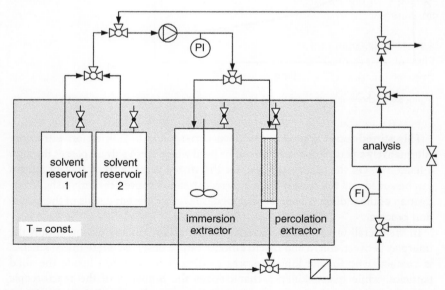

Figure 6.7 Standard laboratory equipment for determining extraction kinetics for percolation and immersion extraction.

With this equipment immersion as well as percolation experiments can readily be performed.

To determine the overall content of solute in the plant material Soxhlet extraction is generally advised. It should be mentioned though that the result of the Soxhlet extraction does not necessarily directly correspond to the true content of the solute in the solid. Apparently there is often a considerable amount of solute trapped either by chemisorption or in the plant-specific structures of the solid that strongly hinder extraction. Only if a condition like temperature is changed severely, can an impression on the overall solute content be obtained. Similarly changing the particle size strongly by intensive grinding may significantly influence the results.

6.3.4
Example Extraction of Pepper

Black pepper (*Piper nigrum*) can serve as a good test example for extraction kinetics. The major transfer component of black pepper is piperine with a content of 5–9%. Piperine is the alkaloid responsible for the pungency of black pepper, along with chavicine, which is an isomer of piperine. Pepper has also been used in some forms of traditional medicine and its ingredient piperine as an insecticide. Piperine extraction from black pepper is especially suited as example case, since piperine can easily be analyzed with UV-Vis spectroscopy and the extraction can be performed with water or ethanol or mixtures of both, where the extraction kinetics and equilibria strongly depend on the ethanol content of the solvent. Thus for the extraction of piperine from black pepper some basic steps are presented here as an example.

6.3.4.1 Calibration with Pure Component
First of all an analytical method should be established. Due to the aromatic rings of piperine UV-Vis spectroscopy can easily be performed with high accuracy. Here a wavelength of 341 nm is chosen. Since pure piperine can be purchased from appropriate suppliers, even quantitative calibration is possible. In Figure 6.8 the calibration curve for piperine is shown for a mixture of ethanol and water with mass fractions of 0.5 each. Piperine was obtained with a purity of at least 98%. The fitted linear curve was obtained from the experimental calibration data.

6.3.4.2 Pre-treatment of Solid Raw Material
In general pre-treatment of the solids is recommended with the aim of reducing mass-transfer resistance. Often grinding is used as pre-treatment to reduce particle size. For the example extraction of black pepper whole grains, grains cut into two as well as ground black pepper have been used for extraction as shown in Figure 6.9. First of all it is seen that smaller particles are extracted significantly faster. Since whole grains of pepper are surrounded by a skin, they are extracted especially slowly. The black pepper ground with an ordinary pepper mill down to

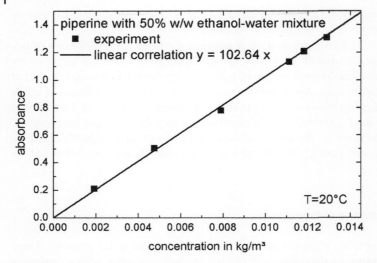

Figure 6.8 Calibration curve of piperine in a mixed solvent with 50 % by mass of ethanol and water at 20 °C.

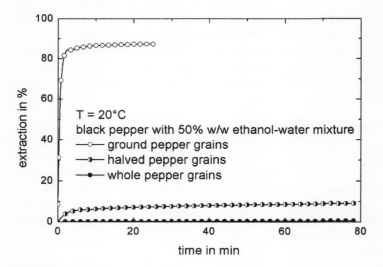

Figure 6.9 Effect of particle size on extraction kinetics.

a typical particle diameter of 0.145 mm is completely extracted in as little as 10 minutes.

6.3.4.3 Extraction Kinetics

The kinetics of the extraction of black pepper with a mixture of ethanol and water with a mass fraction of 0.5 is very fast as shown quantitatively in Figure 6.10.

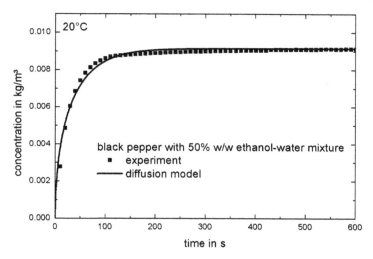

Figure 6.10 Extraction kinetics of ground black pepper with a mixture of ethanol and water with a mass fraction of 0.5 each at 20 °C.

Equilibrium concentration can be approached within roughly 5 minutes for ground pepper. In general for any new system it is recommended that a considerable time should elapse before the kinetics experiment is terminated. In some cases slow and fast extraction kinetics are superimposed and the slow kinetics can only be recognized if the experiment is followed for sufficient time. Here the extraction kinetics is modeled with a simple diffusion model, in which the internal mass transfer is regarded by an effective diffusion coefficient as described in detail in Section 6.4. The initial concentration of the piperine in the particle and its diffusion coefficient are parameters that are fitted to the experimental data obtained in the discontinuously stirred vessel of the standard laboratory equipment. The following values are obtained:

$$D_{\text{eff}} = 0.93 \times 10^{-10} \ (\text{m}^2 \ \text{s}^{-1}) \tag{6.1}$$

$$C_{s,0} = 79.78 \ (\text{kg m}^{-3}) \tag{6.2}$$

where D_{eff} is the effective diffusion coefficient and $C_{s,0}$ is the initial concentration of solute in the solid phase.

6.4
Theoretical Modeling and Scale-Up

Modeling the extraction process of natural raw materials is essential for the design and optimizing of the extraction process including the major process variables. A reliable description of any extraction process should account for mass transfer inside and around the solid raw material.

The overall model can be subdivided into a macro-model describing mass-fluxes on the equipment level and a micro-model describing mass transfer inside and around the solid particles. Thus the micro-model describes how the solvent enters into the solid particles, dissolves and extracts the desired components and releases these components into the surrounding solvent. This is described as a function for example, of the concentration of the transfer components already present in the solvent as it passes the solid as well as possibly other influencing parameters. The macro-model takes this local flux of transfer component into the solvent as input and combines the local fluxes into an overall equipment model.

6.4.1
Macro-Model

The macro-model, which accounts for the overall fluid dynamics of the equipment, is mainly influenced by the type and operation condition of the extractor. These macro-models have an impact on the extraction curves as discussed previously in Section 6.3.2 and consist of mass-balance equations for the solute in the fluid phase. By integration of these differential equations, time-dependent concentration profiles in the liquid are obtained and the extraction curve can be constructed. For the basic types of equipment and the cells of the standard laboratory apparatus, the ideal models are presented in the following. In technical processes deviations from ideality have to be taken into account, e.g. by applying residence-time distributions, dispersion coefficients or by subdividing the equipment into different subcompartments that behave differently, e.g. with respect to local solvent flux or concentration.

6.4.1.1 Ideal Stirred Tank
An ideally stirred tank operated in batch mode corresponds to the immersion principle. The flux of the solute, that is, the transfer component from the solid into the solvent is given as $\dot{m}'''_{\text{solute}}$, the volume specific mass rate of solute extracted from the solid.

The time dependent mass concentration of the solvent C_L inside a stirred tank (Figure 6.11) is obtained from a mass balance.

$$V_L \frac{\partial C_L}{\partial t} = \dot{m}'''_{\text{solute}} V_S + \dot{V} C_{L,\text{in}} - \dot{V} C_L \tag{6.3}$$

where V_L is the volume of the liquid phase and $\dot{V}$ the volume flow rate through the extractor.

For a discontinuously stirred tank (Figure 6.11a) the mass flow of solute in the extractor $\dot{V} C_{L,\text{in}}$ and out of the extractor $\dot{V} C_L$ is zero. If the solvent flows continuously through the equipment (Figure 6.11b), then the mass flow through the continuously stirred vessel must be taken into consideration. When pure solvent is used at the inlet, the mass flow of solute to the extractor $\dot{V} C_{L,\text{in}}$ is also zero.

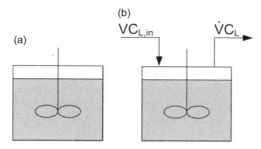

Figure 6.11 (a) Stirred tank batch mode and (b) continuous mode.

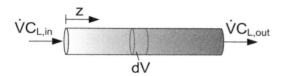

Figure 6.12 Plug-flow extractor of length L.

6.4.1.2 Ideal Plug-Flow Extractor

The plug-flow extractor corresponds to the percolation principle. In a similar way to the continuously stirred tank, the solvent flows through the plug-flow extractor continuously; the extraction curves correspondingly have a similar overall appearance as for the continuously stirred vessel.

To model the plug-flow extractor a one-dimensional model along the axial coordinate z (Figure 6.12) can be set up, where here steady state is assumed and axial dispersion is neglected.

The local concentration in the liquid phase is obtained by

$$\frac{\dot{V}}{A}\frac{\partial C_L}{\partial z} = \dot{m}'''_{solute}(\varepsilon) \tag{6.4}$$

where A is the constant cross-section area, which is assumed as constant and $\dot{m}'''_{solute}$ is depending on the void fraction ε of the solid in the extractor.

6.4.1.3 Other Macro-Models

If the geometry and flow regime in an equipment deviates from these ideal cases, appropriate macro-models should be chosen. It is for example, often helpful to use either a cascade of stirred vessel or a dispersion model to account for deviations from ideality.

For a screw extractor, where co-current and counter-current can be considered, different authors have for example, proposed a model combining tubular reactors and a cascade of continuously stirred tank reactors [6] or a model coupling differential mass balances for the solid and the liquid phase [7].

6.4.2
Micro-Models

For the mass transfer in and around the particles, different approaches can be taken, especially depending on the structure of the natural raw material, determining the mass-transfer rate. A variety of micro-models has been proposed in the literature. In the following the basic features are discussed, then a simple diffusion model is presented, followed by a discussion of some extensions proposed in the literature.

For the extractive separation of the target compound, the main structure of a natural cell is of interest (Figure 6.13). The whole cell is surrounded and protected by the cell wall. The cell wall is also responsible for the more or less rigid structure of the cell. The internal part of the cell includes, among other contents, the vacuole, a large water-filled volume enclosed by a membrane known as the tonoplast. Often the vacuole can be regarded as the main storage place for the solute. To extract the desired component from such a cell structure, the main transport resistances lie within the cell wall and the tonoplast. Another locus relevant for the mass transfer is the intercellular space. In dried plant material the intercellular space can be seen as gaseous void between the neighboring cell walls. Soy beans for example show 20% of intercellular space [8]. These voids can be seen as transport paths through the solid particle.

This biological background shows that the mass-transfer process in extraction from natural raw materials can be described by the following mass-transfer steps for the solute [9]:

1) desorption from the solid matrix
2) diffusive transport within the particles
3) transfer from particle surfaces into the bulk fluid across the liquid boundary layer

In a variety of publications the first step is combined with the second into an intraparticle diffusion step, which is quantified by an effective diffusivity. Step 3

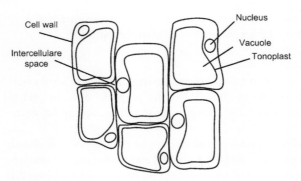

Figure 6.13 Diagram of sugar cane cells [8].

can often be neglected because the rate-determining steps are those inside the particle. For a constant effective diffusion coefficient D_{eff} mass transfer inside the particle can be described by Fick's second law:

$$\frac{\partial C_S}{\partial t} = D_{eff} \left[\frac{\partial^2 C_S}{\partial r^2} + \frac{2}{r} \frac{\partial C_S}{\partial r} \right] \tag{6.5}$$

where C_S is the concentration of the solute within the solid phase depending on time and locus.

In the derivation of this mass-transfer equation the following assumptions were made:

- the particles are spherical, symmetrical and homogeneous,
- the effective diffusion coefficient is constant over time and space,
- the liquid phase in the extractor is perfectly mixed,
- no diffusion boundary layer on particle surface,
- solid particles are completely leached at equilibrium.

The differential Equation 6.5 can be solved, if the starting and boundary conditions are specified. For the typical case the following can be assumed:

- the initial solute concentration in the solid phase $C_{S,0}$ is a given constant, i.e., for $t = 0 : C_S = C_{S,0}$,
- the spherical particle is symmetrical with respect to its center, i.e. a suitable boundary condition is for $r = 0$ and $t > 0$: $\frac{\partial C_S}{\partial r} = 0$.

The conditions at the boundary $r = R$, are dependent on the characteristics of the extraction system. If the external mass-transfer resistance is negligible the analytical equation takes the form [10, 11]:

$$\frac{V_L}{V_S} \frac{C_L(t)}{C_{S,0}} = 1 - \frac{6}{\pi^2} \sum_{n=1}^{\infty} \frac{1}{n^2} \exp\left(-\left(\frac{n\pi}{R}\right)^2 D_{eff} t\right) \tag{6.6}$$

where R is the radius of the solid particle and C_L is the concentration in the liquid phase dependent on time, Equation 6.6 reduces to

$$\dot{m}'''_{solute} = -\frac{3}{R} D_{eff} \frac{\partial C_S(r=R)}{\partial r} \tag{6.7}$$

The effective diffusion coefficient D_{eff} and the initial solute concentration in the solid phase $C_{S,0}$ can be fitted to experimental data.

6.4.2.1 Other Micro-Models

Analytical models largely simplify the estimation of model parameters from the shape of experimental extraction curve [4]. Thus it is a common practice to keep the mass-balance equations discrete not only in time but also in space and then solve the balances numerically.

As a result of plant pre-treatment by grinding, milling, cutting or other methods of comminution, the particles are rarely uniform, so that different shapes, sizes and surface areas are obtained. For calculation of the mass-transfer coefficient, therefore, different particle geometries can be accounted for, mainly spheres, cylinders and slabs [12]. Also particle-size distributions can be taken into account, that generally lead to a fast mass-transfer kinetics at the beginning and a slower kinetics at the end of the extraction process.

In order to describe the nonhomogeneous structure of the solids, several suggestions have been made. Often a fast extraction of solute readily accessible on the surface of the cells must be accounted for as has been found in vanilla. After this first fast extraction step, the rest is extracted according to the effective-diffusion model. In other cases the solid may have a solubility limit that must be overcome in the beginning. This leads to the shrinking-core model [13], which is widely used. The desorption equilibrium can be described with corresponding models in this case [1]. The particle is modeled as an inner zone, the core, and an outer zone. Within the core the solubility limit of the solute is exceeded, thus a constant concentration corresponding to this limit is reached. This core shrinks over time as a consequence of the transfer of the solute out of the particle. In the outer zone mass transfer can, for example, be described by effective diffusion.

Another approach to model the internal mass transfer is the concept of broken and intact cells. This approach takes into account the sudden reduction in extraction rate that is observed after the first extraction period, particularly when vegetable oil is extracted from seed [4]. Two regions are distinguishable in the particle. Close to the surface there is a region of broken cells, whose walls have been damaged by mechanical pre-treatment, and the particle core, that contains intact cells. The mass-transfer resistance of cell walls is high and, therefore, there is a large difference in the diffusion rates of both regions. The initial fast extraction from broken cells is followed by a much slower extraction from intact cells.

Also the internal structure of the individual cell types can be considered in more detail, e. g. for the extraction of essential oils. A model has been proposed by Zizovic [14], which is based on the hypothesis that the oil extraction should be dependent on the type of secretory ducts that are the oil reservoirs of the plant. They determine the internal structure of the plant by SEM micrographs and integrate the effect into a time-dependent variable, the so called "Source and Transfer" term [14].

To extend models, the different effects of mass transfer described above, like the shrinking-core and the method of broken and intact cells [15], can be combined.

Finally, there are also empirical models for the extraction of natural raw materials [16], like the standard function method [17]. This model consists of an exponential function with three parameters A, B, H of the form:

$$C_L = A - B * exp(-Ht) \tag{6.8}$$

where the parameter A approximates to the equilibrium concentration $C_{L,eq}$ in the liquid phase.

Since the scale-up of an extraction process depends on reliable modeling, a suitable micro-model needs to be chosen. It has been shown that the significance of model discrimination can be increased, when the micro-models are used to evaluate experimental results for different extractor types e. g. in the standard laboratory experiments. Due to the different principle shape of the extraction curve depending on extractor type, discrimination of the best micro-models can be performed more precisely [5].

6.4.3
Fitting to Experimental Data and Model-Based Experimental Design

Most of the models for the description of mass transfer include parameters that need to be fitted to experimental data. A curve-fitting technique permits the determination of parameter estimates and their confidence interval from a data set consisting of experimental data points [18]. Since the performance of the fitting procedure is strongly dependent on the starting values for the parameters, preliminary knowledge of the range of the parameters is helpful. The number of data points considered should not be below roughly double the number of fitted parameters as a rule of thumb.

Since data are often obtained for a time series of concentration, the obtained data are not independent. Thus it is likely that highly correlated parameters result. In such a case model-based experimental design (or model-based experimental analysis, MEXA) can be used to reduce the correlation between the parameters or to reduce the experimental effort considerably to significantly perform model discrimination and parameter estimation [19]. In MEXA the experimental conditions are chosen in such a way as to maximize the gain in significance of the model or model parameters for each experiment. Thus the information obtained by each experiment is maximized and the number of experiments can be minimized without reducing the information achieved. This information is generally measured with the Fisher information matrix. The details of MEXA are described in detail elsewhere [20–22]. Typically a reduction in experimental effort of factor 3 to 10 can be achieved [23]. The approach of model-based experimental design is especially promising, if the number of influencing parameters is high and a variety of comparably similar models is to be discriminated. Since this is often the case in extracting plant material, model-based experimental design appears to be most suited also for the design of experiments to characterize the kinetics of solid extraction.

6.4.4
Scale-Up Design

For the design and scale-up of technical equipment, laboratory experiments are usually performed to obtain quantitative information on the content of the desired components in the solid and to get a first impression on extraction kinetics. The latter is often determined in a simple beaker experiment or with a small percolator resulting in an extraction curve. In this experiment also the influence of the major process parameters is quantified, like effects of temperature, solvent type and concentration. The scale-up is generally performed empirically and based on experience. Then typically small pilot-plant scale experiments are performed for validation and further optimization of the process conditions.

6.5
Industrial Extraction Equipment

6.5.1
Pre-processing of Plant Material

Before a plant material is subjected to solvent extraction, typically several pre-processing steps are applied. These comprise: drying, thermal or enzymatic pre-treatment, pressing, grinding, milling, extrusion, micronization, pelletization, granulation, swelling, pre-wetting, and others.

A drying step is normally required to avoid microbiological contamination of the plant material during storage. At the same time the total weight of the product is decreased, resulting in reduced transportation costs. A thermal pre-treatment can also serve to coagulate or denature proteins, inactivate enzymes or to generate stable matrices, which are better to handle in industrial extraction equipment. The selection of drying equipment depends mainly on the mechanical properties and thermal sensitivity of the raw material and the quantities to be handled. Equipment for drying is not described here. A comprehensive summary of available drying technologies, equipment, applications and suppliers is given in Refs [24] and [25].

A particle-size reduction by pressing, grinding, milling or extrusion is almost always applied for two reasons: first, the specific weight of the plant material is increased at reduced particle size, leading to reduced transportation costs; second, the extraction efficiency is improved with reduced particle-size. This is because diffusion of the solvent inside the solids is the limiting mass-transfer mechanism. A disruption of the plant cells makes direct extraction into the solvent possible. Reducing the particle size results in shorter diffusion pathways and hence reduced extraction time. However, size reduction of particles is limited due to technical difficulties. Depending on the type of extraction equipment and operation too small particles can lead to a high pressure drop, slow drainage, agglomeration, flow instabilities, channeling, formation of lakes or entrainment of fines into the extract. As a rule of thumb the portion of fine particles with sizes <0.5 mm should not exceed 5 to 10% [26]. Selection of equipment for particle size reduction depends on physical properties and structure of the plant material and the target particle size and structure. For example oil seeds are typically pre-treated in hydraulic presses prior to solvent extraction. Other typical equipment for size reduction of plants are shredders, cutting mills or ball mills. An extensive overview on size reduction technologies is given in Ref [27].

In some cases pelletization or granulation of the plant material to obtain particles of defined size can be advantageous, namely in operations where a high flow rate of the solvent through the solids is required. For pelletization the milled plant material is moisturized with water or steam and subsequently extruded through screens and perforated plates. For size enlargement by granulation different types of equipment like mixers, drum, pan or fluidized-bed granulators are applied. An overview on technologies for size enlargement is given in Ref [28].

Swelling with water to increase the volume of plant material and to improve the permeability of cell walls by induced cell lysis can sometimes improve the extraction. A pre-swelling outside the extractor may be required in some cases to avoid uncontrolled volume increase of the plant material inside the extraction equipment. Less common pre-processing methods are enzymatic pre-treatment to improve permeation of solvents through the cell walls, and the application of compressed gases for swelling or micronization by rapid expansion. The applications of compressed gases for pre-treatment and extraction are described in Chapter 4.

6.5.2
Selection of Extraction Solvents

For the selection of extraction solvents different aspects to be considered are:

- physical and chemical properties,
- availability and economics,
- safety and toxicological aspects,
- regulatory issues.

The most important physical properties are solubility and selectivity of the target components, stability and boiling point of the solvent. Physical and thermodynamic models to predict the solubility of natural compounds are described in Chapter 2 of this book. Besides that viscosity, interfacial tension, heat of vaporization and latent heat have to be considered for solvent selection.

Safety aspects cover flammability (ignition and flashpoint), chemical and thermal stability, toxicity and environmental aspects like biodegradability. The selection of solvents for the extraction of components with certain applications (e.g., foods, cosmetics, pharmaceuticals) must comply with laws and regulations for these product groups (see Chapter 9).

Commonly used solvents for plant extraction are water, alcohols (mainly methanol, ethanol and isopropanol), glycols, hydrocarbons (e.g., hexane, heptane), ketones (e.g., acetone, methyl ethyl ketone), acetates (e.g., ethyl, propyl, butyl acetate) and compressed or supercritical gases (carbon dioxide, propane, butane, nitrous oxide (refer also to Chapter 1)). Recently alternative solvent systems like surfactant-water solutions (see Chapter 3 of this book) and ionic liquids [29] have been suggested, mainly driven by consumer acceptance and environmental aspects.

6.5.3
Solid–Liquid Extraction Processes

The choice of extraction principle for a given extraction problem mainly depends on which is the limiting mechanism in the transfer of the extract component from the solid into the liquid. Four main transport mechanisms can be distinguished:

- convective transport of solvent from and to the bulk of the liquid to or from the particle surface,
- diffusion of solvent and extract through laminar film at the particle surface (film diffusion),
- diffusion of solvent and extract through solid pores (pore diffusion), and
- dissolving of extract component into the solvent.

Percolation is preferable for solid materials with low internal resistance to mass transfer, while immersion is the better choice, when the resistance to diffusion inside the flakes is high. For percolation the solid material must be modified to achieve a high flowrate of the solvent through the solid bed. For example, oil seeds are pressed to flakes, fine powders are pelletized to allow for a high percolation velocity, which is beneficial for washing away the extract solution from the particle surface. For immersion the particle size should be decreased to reduce the diffusion paths inside of the solids. Immersion is applicable also to solids, which tend to swell or to disintegrate. A disadvantage of immersion compared with percolation is the need for mechanical separation of the solids after extraction, either by filtration or centrifugation. This is the reason that the percolation principle is preferred over immersion technologies for most industrial-scale operations. Parameters influencing extraction efficiency are solvent composition, temperature, particle size, liquid-to-solid ratio and pH value. Aspects for selection of the most suitable operation mode (single batch, continuous multistage, countercurrent extraction) are discussed in the next sections.

6.5.3.1 Batch Processes

The simplest mode of SLE is the operation in a single batch. This can either be done as immersion extraction in a stirred vessel or centrifugal extractor or as percolation extraction in a vessel with filtering bottom [30]. In the first case the extraction is finalized, when the equilibrium of the extract concentration between the solids and the liquid is reached, meaning that complete extraction is not achieved. For a complete extraction the repeated addition of fresh solvent is required. However with proceeding depletion of the plant material the extract concentration in the solvent becomes lower and lower, resulting in high overall solvent consumption.

In percolation processes fresh solvent is continuously trickled through the solid bed, so that a good concentration difference of the extract between solids and liquid is always maintained. Nevertheless with proceeding depletion of the solids the concentration of the extract in the solvent decreases and the solvent consumption is comparable high like in the single batch immersion.

Another mode of batch extraction is the application of the Soxhlet principle. Solvent is circulated by percolation through a bed of the plant material and the rich miscella is subsequently concentrated in an external evaporator to recover the solvent. The solvent is recycled to the extractor, while the concentrate is continuously taken from the evaporator. This procedure has the advantage that always the highest possible concentration difference is maintained and thus a complete extraction is achieved. The disadvantage is still the high solvent consumption.

Due to their high solvent consumption, single-batch extractors have almost completely disappeared and are now only used for small product rates (e.g., lavender, flavors) [26]. An improved procedure is the combination of several batch extractors in a multistage operation [30]. The most common is the application of this procedure in industrial scale for a series of batch percolators, as illustrated in Figure 6.14.

At the beginning fresh solvent is fed to the first percolator and extract enriched miscella leaving this percolator is fed to the miscella tank. When the extract concentration in the effluent solvent is decreasing due to the proceeding depletion of the solids, the partly enriched solvent is fed to the second percolator, where additional extract is taken up by the solvent. The effluent solvent from the second percolator is then led to the miscella tank as long as the extract concentration is high enough. If the extract concentration falls below a defined value the partly enriched solvent leaving the second percolator is fed to the third percolator to further increase the extract concentration. The same procedure is subsequently applied to all percolators. When the solids in the first percolator are completely depleted, the fresh feed is switched to the second percolator and the first percolator is desolventized and unloaded from the depleted plant. After that the percolator

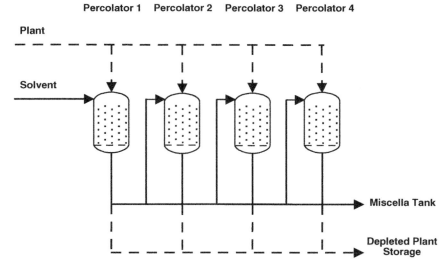

Figure 6.14 Four stage batch percolation plant.

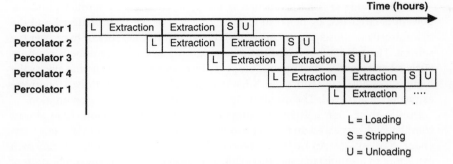

Figure 6.15 Batch-cycle diagram for a four-stage percolation procedure.

is refilled with fresh plant material and is again connected into the multi-batch extraction operation. A typical percolation battery consists of three to six percolators in series.

Figure 6.15 shows a typical batch-cycle diagram for a four-stage percolation battery. With the described procedure much higher extract concentrations are achieved resulting in substantial reduction of the solvent requirement and hence lower energy consumption in the subsequent solvent-recovery step.

The multistage procedure can be modified in different ways to further optimize extraction time and solvent usage. For example solvents in a first step can be recirculated through the solids bed to increase extract concentration more quickly before leading the miscella to the extract tank. Another option is to combine more than two percolators during the extraction phase to further increase the extract concentration. The optimum procedure is dependent on the extraction behavior of each plant material and should be evaluated by laboratory or pilot trials.

6.5.3.2 Continuous Processes

A further improvement of industrial extractions with regard to extraction efficiency, solvent and energy consumption and convenience in operation is achieved by the application of fully counter-currently operating processes. Several different types of counter-currently working industrial extraction equipment have been developed for different applications. The available equipment can be classified as either immersion or percolation-type equipment. The selection of the proper type of equipment for a given extraction problem depends on the one hand on the physical properties of the raw materials to be extracted, and on the other hand on the quantities of raw materials which have to be handled in a given time. For example for the extraction of herbs or very valuable phytoextracts (e.g., for pharmaceutical applications), the extraction of small quantities of plants, between several 100 kg and a few tons per day, is typical, while the biggest extractors for oilseed processing work with up to 10 000 tons of raw material per day. Other aspects for selection of the proper extraction equipment are single versus multiple product operation, frequency of product changeover and labor versus capital costs.

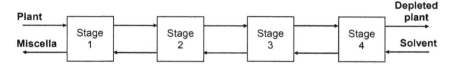

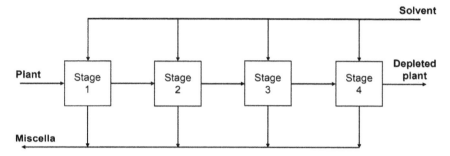

Figure 6.16 Counter-current and cross-current solid liquid extraction.

Some of the extraction equipment described in the next section works in the fully counter-current mode, which means, that plant and solvent run counter-currently from stage to stage through the whole equipment (Figure 6.16). The advantage of this procedure is that the fresh solvent is brought into contact with the most depleted plant, while the fresh plant material is contacted with already enriched miscella. Thus a high concentration difference of the extract components between the solids and the liquid is always achieved. Another option is to run the extraction in cross-current mode, meaning that the solids are transferred from stage to stage through the whole equipment, while fresh solvent is introduced into each extraction stage (Figure 6.16). This operation mode favors complete extraction of the value material, because the highest possible concentration difference is always maintained. However the required amount of solvent is higher than in the case of the counter-current extraction. In most of the commercially available equipment both operation procedures are combined, in such a way that in certain sections of the extractor, solids and liquids run counter-currently to enrich the miscella, while in another section the solids are washed with fresh solvent in cross-current mode, to achieve a complete depletion of the plant material.

6.5.4
Extractor Design

The design of the extraction apparatus differs according to the application of the plant-extraction technology. Food processing like sugar production, oil separation or coffee extraction requires continuous processing with high throughput of raw material. On the contrary the production of pharmaceuticals needs smaller but more flexible apparatus and process design. About 20 companies manufacture industrial extractors worldwide. Some examples for typical designs are shown in the following.

6.5.4.1 Batch Equipment

Batch extraction can be done in a simple vessel or with solvent recycle (Soxhlet type) (Figure 6.17) or a stirred apparatus as immersion extraction or in an apparatus with filter or sieve plate at the bottom as percolation extraction. Today the type of simple static vessels is substituted by equipment in which solvent and material are in movement.

Immersion Extraction Examples of this equipment with mixing devices are the Lödige plugshare mixer (Figure 6.18) and centrifuges or decanters like the GEA Westfalia Separator (Figure 6.19) [34] and the Graesser-Contactor that can be used batchwise as well as continuously. These extractors are mostly used in mixing processes without mass transfer and also in reaction, but they are adapted to SLE.

It is well known from theory and experience with all kinds of material that diffusion is not the only important factor for mass transfer, but mostly the intensive

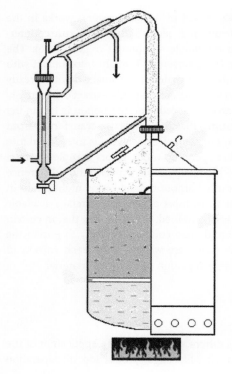

Figure 6.17 Soxhlet-extractor (Courtesy of Albrigi Luigi s.r.l.) [31].

Figure 6.18 Lödige Plowshare mixer (Courtesy of Gebr. Lödige Maschinenbau GmbH) [32].

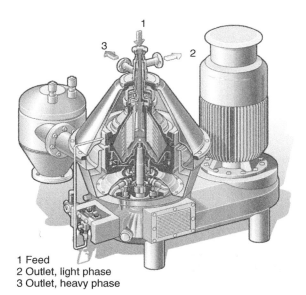

1 Feed
2 Outlet, light phase
3 Outlet, heavy phase

Figure 6.19 GEA Westfalia Separator (Courtesy of GEA Niro (Denmark)) [33].

Figure 6.20 Percolation extractor 1 m³; (Courtesy of e&e Verfahrenstechnik GmbH) [35].

solid–liquid contact is responsible for the mass transfer from solid into the bulk of liquid. And in addition the driving force, the concentration difference between the solid surface and the surrounding liquid, is increased by good mixing.

Percolation Extraction In percolation extractors, the solids are in the form of a fixed bed and the solvent is pumped through or sprayed over the material. The batch type of percolation extractor is mostly designed in a conical shape closed at the bottom by a filter and a valve. The construction has to be such that the times for the filling and discharging process, which are sometimes automated by pneumatic or hydraulic equipment, are reduced. The solids must be coarsely ground to allow a good drainage through the bed. Because of the swelling of the organic material the solids must be moistened before filling. A serial of apparatuses are used to reduce solvent consumption as mentioned before. A technical example of this batch percolation is shown in Figure 6.20. Batch extraction is widely used for the so-called standard extracts which are produced from many different plants.

6.5.4.2 Continuous Extractors

Batch percolators are more and more replaced by continuously and countercurrently operated apparatuses [26, 34, 36–38]. This depends mainly on the

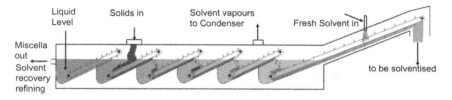

Figure 6.21 Crown Iron Model IV (Courtesy of Crown Iron Works Company) [39].

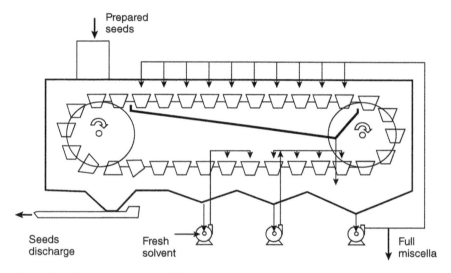

Figure 6.22 Basket-type extractor [40].

quantity of raw material which is to be extracted. Most of the continuous apparatuses operate as percolation type but there is also an immersion type, the Crown Iron model IV (Figure 6.21). The raw material is transported by belts countercurrently to the solvent and dips into the solvent during transportation.

One of the first continuous extractors was the basket-type extractor (Figure 6.22) wherein the material is carried in baskets supported by endless chains. It was built as a horizontal and as a vertical type. This type was later improved in form of the Lurgi sliding cell extractor (Figure 6.23). Because of mechanical problems it was followed by rotary (carrousel) extractors (Figures 6.24 and 6.25), mostly used for soy bean oil extraction, belt extractors (Figure 6.26), screw extractors (Figure 6.27) and U-type extractors wherein sieves are arranged on a conveyor belt. Many of these types were developed for the oil or sugar industry. There are three designs of sugar beets extractors: the horizontal rotating "RT" of Raffinerie Tirlemontoise;

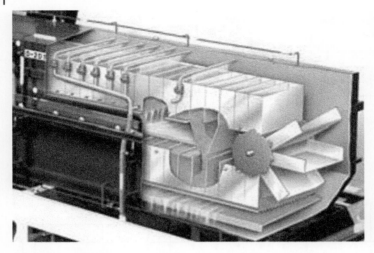

Figure 6.23 Lurgi sliding cell extractor (Courtesy of Lurgi AG) [41].

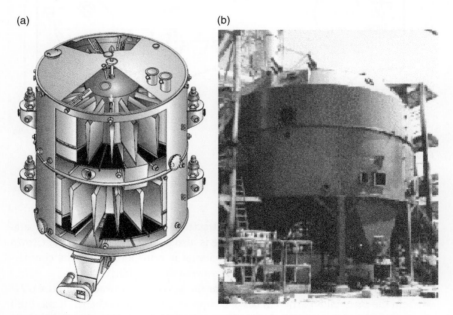

Figure 6.24 Carrousel extractor, (a) Harburg Freudenberger Maschinenbau (Courtesy of Harburg Freudenberger) [42], (b) De Smet Reflex-Extractor (Courtesy of De Smet Ballestra Group N.V.) [43].

6.5 Industrial Extraction Equipment | 211

Figure 6.25 Pruess Rotary extractor (Courtesy of Pruess Anlagentechnik GmbH) [44].

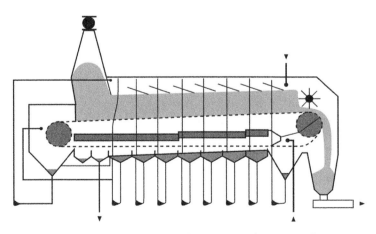

Figure 6.26 De Smet LM-Extractor (Belt) (Courtesy of De Smet Ballestra Group N.V.) [43].

(a)

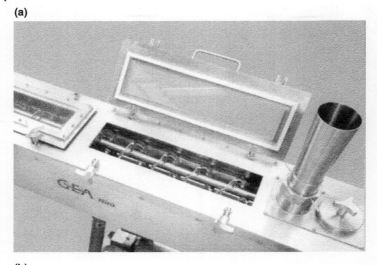

(b)

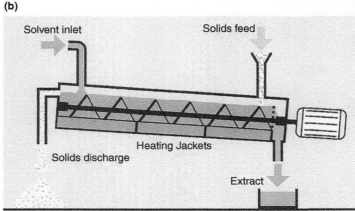

Figure 6.27 GEA Niro Contex (Screw) (Courtesy of GEA Niro (Denmark)) [45].

the inclined screw of "DDS" (De Danske Sukkerfabrikker) (Figure 6.28); and the tower design, developed by BMA, Braunschweigische Maschinenbauanstalt AG (Figure 6.29) and Buckau-Wolf. Inside the tower a shaft rotates slowly. Helicoidal flanges on the shaft transport the sugar beet cossettes upwards. The sugar juice and the cossettes move counter-currently [47].

6.5.5
Future Developments

Plant extraction is a mature technology, where most of the procedures and equipment used today were developed almost hundred years ago. In recent years

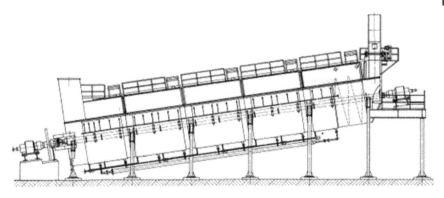

Figure 6.28 DSSE Screw Extractors (Courtesy of Danish Sugar Sweetener Engineering A/S) [46].

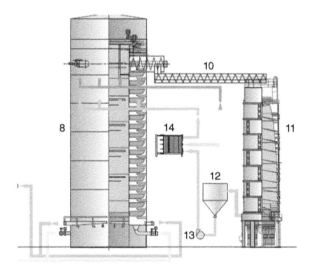

Figure 6.29 BMA tower extractor (Courtesy of Braunschweigische Maschinenbauanstalt AG) [47].

however the interest in plant extraction has grown due to the trend for using renewable resources. This has initiated new research and developments in the field of plant extraction [48, 49]. Reported advanced plant-extraction methods have been mostly applied on the laboratory scale, e.g., for analytical or preparative purposes [30]. These include:

- ultrasound or microwave-assisted extraction (UAE, MAE),
- pressurized fluid extraction (PFE), accelerated solvent extraction (ASE),
- use of alternative solvents.

Ultrasound is applied to improve the mass transfer during extraction. Fundamentally, the effects of ultrasound on the cell walls of plants can be described as follows:

- Some plant cells occur in the form of glands (external or internal) filled with essential oil. A characteristic of external glands is that their skin is very thin and can be easily destroyed by sonication, thus facilitating release of essential oil contents into the extraction solvent.

- Ultrasound can also facilitate the swelling and hydration of plant materials to cause enlargement of the pores of the cell wall. Better swelling will improve the rate of mass transfer and, occasionally, break the cell walls, thus resulting in increased extraction efficiency and/or reduced extraction time.

Moreover the application of ultrasound generates an oscillating pressure swing, which periodically accelerates the flow of solvent inside the solids pores. This effect additionally improves mass transfer.

The use of microwaves has the advantage that solvents are heated quicker than with indirect external heating. Moreover the heating occurs inside the solids, where the dissolution of the extract components takes place. Due to both effects the extraction is accelerated. In the application of MAE the nature of the solvent obviously is of prime importance. In general the solvent should absorb the microwaves without leading to strong heating so as to avoid degradation of the extract components. It is common practice to use a binary mixture (e.g., hexane–acetone, 1+1), with only one solvent absorbing microwaves. In some cases, polar solvents such as water or alcohols may give efficient extractions.

For both UAE and MAE it is questionable, whether these technologies can economically be transferred to the manufacturing scale, since the technical complexity of applying ultrasound or microwaves in large-scale equipment is high. It is conceivable to use it in belt type extractors because of the rather thin layer of solids.

Pressurized fluid extraction (PFE), also referred to as accelerated solvent extraction (ASE), was developed for analytical purposes [48]. It uses elevated temperatures (100 °C to 180 °C) and pressures (100 bar to 140 bar). The idea is, that solvents heated above their boiling point have very good extraction efficiencies. The effects are similar to SFE with the difference, that the latter is limited to a handful of inert gases, while the PFE can be applied to every type of solvent. For analytical purposes extractions with PFE are as efficient as Soxhlet extraction but use much less solvent and take significantly less extraction time [48]. Again the question is, whether this technology is economically transferable to the manufacturing scale.

Application of pressure to improve extraction is also the basis of a patented procedure [50], which is however performed at low temperature and hence is especially useful for thermally sensitive products like herbs, flavors etc. (Figure 6.30). The principle is a batch maceration, in which the vessel is pressurized to 8–10 bar. The pressure improves the diffusion of the solvent into the plant material. At depressurizing the system kind of a suction effect occurs which improves

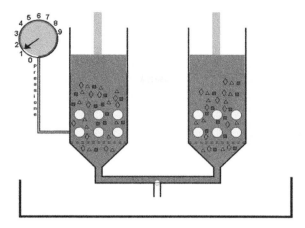

Figure 6.30 Naviglio extractor [50].

the extraction efficiency. Repeating the procedure a number of times is claimed to give a complete depletion of the plant material like in a Soxhlet extraction. The efficiency of this procedure is still under investigation at a university institute and was demonstrated in a 50 l prototype for different plant materials with good results [34].

The application of environmentally friendly solvents for extraction of plant materials is another area of future developments. Besides ionic liquids, supercritical extraction with CO_2 or other gases the application of surfactant/water systems is investigated, which is described in Chapter 3.

6.5.6
Extract Treatment

The rich miscella from plant extraction must be subjected to further treatments before achieving the extract in the desired final form. Typical required treatments are:

- filtration to separate solids,
- concentration of extracts by evaporation of solvents,
- enrichment of target components by liquid-liquid extraction, chromatography, membrane separation, crystallization etc.,
- removal of impurities like heavy metals by complexation, or polyaromatic hydrocarbons by adsorption technologies,
- heat treatment, pasteurization of liquid extracts,
- drying of products and product forming technologies for solid extracts (e.g., spray drying, granulation etc.).

The discussion of downstream processing of plant extracts is not part of this chapter, but described in detail in Chapter 5.

6.5.7
Depleted Plant Treatment

The depleted plant after solvent extraction typically contains between 35 and 50% solvent, which must be recovered for economical and environmental reasons. Possible methods are pressing, washing, heating or steaming. Pressing is not very common, since it is technically complicated. Washing with water is sometimes applied but requires additional treatment of the solvent-water mixtures. Moreover washing alone is normally not efficient to achieve the final low solvent levels required for the disposal of the plant material.

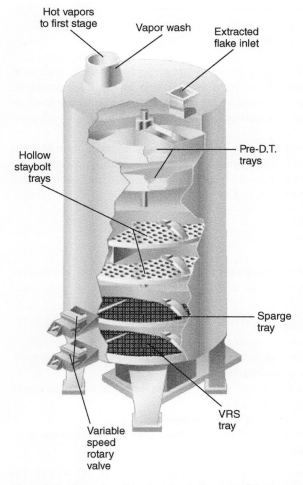

Figure 6.31 Crown Iron "desolventizer" (Courtesy of Crown Iron Works Company) [51].

More common is the recovery of solvents from the depleted plant by heating or steaming. Batch percolators are typically desolventized before unloading the depleted plant material by first pulling vacuum and subsequent introduction of steam at the filtering bottom of the percolator. The solvent-water vapors are condensed and have to be worked up separately by phase separation and/or distillation. Disadvantages of this method are low efficiency with high steam consumption, high final solvent content of the plant material (1–5%) and the entrainment of solids to the vapors, which requires additional filtration of the condensates. Steaming from the top of the percolators has the additional disadvantage, that plant ingredients are extracted with the condensing steam, which subsequently have to be separated from the solvent-water mixture.

Continuously running extractors are typically supplied with a continuous desolventizing unit. Most common is the so called "desolventizer/toaster" supplied for example, by De Smet, Lurgi and Crown Iron [39, 41, 43] in Figure 6.31. The equipment consists of a tower with heated plates. The solvent-wetted plant material is continuously fed to the top of the tower and is transported from top to bottom of the tower by scrapers and open screens on each plate. In the upper part of the tower the plates are indirectly heated, evaporating the major part of the solvent from the plant material. In the lower section of the tower life steam can additionally be introduced to improve the solvent recovery and to achieve the required low solvent level in the depleted plant. However, depending on the applied solvents, in modern designs the addition of steam is avoided, to circumvent additional treatment of solvent/water mixtures.

If the depleted plant material after desolventation meets the environmental regulations, it is typically used by farmers for composting. If the solvent levels are too high, the use of the material as fuel or for biogas generation can be considered.

Symbols

Latin Symbols

A area [m^2]
C concentration [kg m^{-3}]
D_{eff} effective diffusion coefficient [m^2 s^{-1}]
$\dot{m}'''_{solute}$ volume specific mass rate of solute extracted from the solid [kgm^{-3} s^{-1}]
r radial coordinate [m]
R radius [m]
t time [s]
V volume [m^3]
$\dot{V}$ volume flow [m^3 s^{-1}]
z axial coordinate [m]

Greek Symbols and Special Characters

ε void fraction [-]

Subscripts

0   initial
in   input
L   liquid phase
S   solid phase
out   output

References

1 Stockfleth, R. and Fiebig, B. (2004) *Dtsch. Lebensmitt. Rundsch.*, **100** (9), 343–347.
2 Sattler, K. (2001) *Thermische Trennverfahren – Grundlagen, Auslegung, Apparate*, 3rd edn, Wiley-VCH Verlag GmbH, Weinheim, Germany.
3 Sticher, O. (2008) *Nat. Prod. Rep.*, **25**, 517–554.
4 Sovová, H. (2005) *J. Supercrit. Fluids*, **33**, 35–52.
5 Delinski, D. and Pfennig, A. (2007) Standardisierung einer Laborapparatur zur quantitativen Vermessung der Phytoextraktionskinetik. ProcessNet, Asselheim.
6 Prat, L., Guiraud, P., Rigal, L. and Gourdon, C. (1999) *Chem. Eng. Process.*, **38**, 73–83.
7 Simeonov, E., Seikova, I., Pentchev, I. and Mintchev, A. (2003) *Ind. Eng. Chem. Res.*, **42**, 1433–1438.
8 Schneider, F.H. (1980) *Extraktive Trennung Fest/Flüssig – Untersuchungen über die Feinstruktur vegetabiler Feststoffe und ihren Einfluss auf das Extraktionsverhalten*, Westdeutscher Verlag, Opladen, Germany.
9 Winitsorn, A. and Douglas, P.L. (2008) *Chem. Eng. Commun.*, **195**, 1457–1464.
10 Henschke, M. (2003) *Auslegung Pulsierter Siebboden-Extraktionskolonnen*, Shaker Verlag, Aachen, Germany.
11 Crank, J. (1975) *The Mathematics of Diffusion*, 2nd edn, Clarendon Press, Oxford.
12 Reverchon, E. (1996) *AIChE J.*, **42** (6), 1765–1177.
13 Goto, M. and Roy, B.C. (1996) *J. Supercrit. Fluids*, **9**, 128–133.
14 Zizovic, I. and Stamenić, M. (2007) *J. Supercrit. Fluids*, **39**, 338–346.
15 Fiori, L. and Basso, D. (2009) *J. Supercrit. Fluids*, **48**, 131–138.
16 Cocero, M.J. and García, J. (2001) *J. Supercrit. Fluids*, **20**, 245–255.
17 Simeonov, E., Tsibranska, I. and Minchev, A. (1999) *Chem. Eng. J.*, **73**, 255–259.
18 Toledo, R.T. (2007) *Fundamentals of Food Process Engineering*, 3rd edn, Springer, New York.
19 Babic, D. and Pfennig, A. (2006) *Fluid Phase Equilibr.*, **245**(2), 140–148.
20 Asprey, S.P. and Macchietto, S. (2000) *Comp. Chem. Eng.*, 24, 1261–1267.
21 Asprey, S.P. and Macchietto, S. (2002) *J. Process Control*, **12**, 545–556.
22 Kalem, M., Bertakis, E. and Pfennig, A. (2008) *Chem. Ing. Tech.*, **80** (1–2), 79–85.
23 Bertakis, E., Kalem, M. and Pfennig, A. (2008) *Chem. Eng. Sci.*, **63**, 4481–4887.
24 Tsotsas, E., Gnielinski, V. and Schlünder, E.-U. (1988) Drying of Solids, in *Ullmann's Encyclopedia of Industrial Chemistry*, 5th edn, vol. B2, (ed. W. Gerhartz), Ch 4, Wiley-VCH Verlag GmbH, Weinheim, Germany.
25 GVC Fachausschuss Trocknungstechnik, Portal für Industrielle

Trocknungsanlagen, www.drying.de (accessed 1 June 2009).
26. Eggers, R. and Jaeger, P.T. (2003) Extraction Systems, in *Extraction Optimization in Food Engineering* (eds C. Tzia and G. Liadakis), Dekker, New York, Ch 4.
27. Bernotat, S. and Schönert, K. (1988) Size Reduction in *Ullmann's Encyclopedia of Industrial Chemistry*, 5th edn, vol. B2, (ed. W. Gerhartz), Ch 5, Wiley-VCH Verlag GmbH, Weinheim, Germany.
28. Sommer, K. (1988) Size Enlargement in *Ullmann's Encyclopedia of Industrial Chemistry*, 5th edn, vol. B2, (ed. W. Gerhartz), Ch 7, Wiley-VCH Verlag GmbH, Weinheim, Germany.
29. Toledo, R.T. and Kurzhals, H.-A. (2004) *Verfahrenstechnische Grundlagen Der Lebensmittelproduktion*, 1st edn, Behr, Hamburg, Germany.
30. Voeste, T., Weber, K., Hiskey, B. and Brunner, G. (1988) Liquid-Solid Extraction, *Ullmann's Encyclopedia of Industrial Chemistry*, 5th edn, vol. B3, (ed. W. Gerhartz), Ch 7, Wiley-VCH Verlag GmbH, Weinheim, Germany.
31. Albrigi Luigi srl, Standard equipment for the processing of extracting ... http://www.albrigiluigi.com/impianti_erboristeria.html (accessed 28 November 2010).
32. Loedige, Process Technology, Ploughshare-Mixer, http://www.loedige.de/Ploughshare®%20-%20Batch%20Mixer.html (accessed 28 November 2010).
33. GEA Westfalia Separator Group, Separator designs, http://www.westfalia-separator.com/products/separators.html (accessed 28 November 2010).
34. List, P.H. and Schmidt, P.C. (1989) *Phytopharmaceutical Technology*, CRC Press, Boca Raton, FL, USA.
35. e&e Verfahrenstechnik GmbH, Extraction plants, http://www.eunde-verfahrenstechnik.de/extraction.html (accessed 28 November 2010).
36. Bombardelli, E. (1991) Technology for the Processing of Medicinal Plants, in *The Medicinal Plant Industry* (ed. R.O.B. Wijesekera), CRC Press, Boca Raton, FL, USA, Ch 7 pp. 85–98.
37. Tzia, C. and Liadakis, G. (eds) (2003) *Extraction Optimization in Food Engineering*, Marcel Dekker, New York, USA. (Food Science and Technology, 128).
38. Gertenbach, D. (2007) Solid–liquid Extraction Technologies for the Manufacturing Nutraceuticals, Chapter 11 in *Functional Food Ingredients and Nutraceuticals. Processing and Technologies* (ed. J. Shi), CRC/Taylor & Francis, Boca Raton, FL, USA, p. 331ff.
39. Crown Iron Works Company, Model IV Extractor, Diagram, http://www.crowniron.com/technologies/spx_model4.cfm (accessed 28 November 2010).
40. Gümüskesen, A. S. and Yemiscioglu, F. Basket-type extractor, www.food.ege.edu.tr (accessed 1 June 2009).
41. Lurgi AG, Solvent Extraction from Oilseeds, http://www.lurgi.com/website/fileadmin/pdfs/brochures/Br_solventExtr.pdf (accessed 28 November 2010).
42. Harburg-Freudenberger Maschinenbau GmbH, Speiseöltechnik, Extraktion, Karussell-Extraktor, http://www.harburg-freudenberger.com/files/prospekt_extraktion.pdf (accessed 28 November 2010).
43. De Smet Ballestra Group N.V., Extraction, Reflex Extractor, http://www.desmetgroup.com/extraction.html#Anchor-33869 (accessed 28 November 2010).
44. Pruess Anlagentechnik GmbH, Continuous Solvent Extraction, Pruess Rotary Extractor, http://www.pruess.net/index2.htm (accessed 28 November 2010).
45. GEA Niro A/S, Powder Technology Division, Continuous Extraction, http://www.niro.com/niro/cmsresources.nsf/filenames/776Cont_Extractraction.pdf/$file/776Cont_Extractraction.pdf (accessed 28 November 2010).
46. DSSE Danish Sugar Sweetener Engineering A/S, Sugar Equipment, DSSE Diffusers, http://www.dsse.biz/index.php (accessed 1 June 2009).

47 BMA Braunschweigische Maschinenbauanstalt AG, Beet Extraction Plant, Extraction Tower, http://www.bma-de.com/fileadmin/Templates/BMA/pdf/gb/Prospekte/Extraktion_en.pdf (accessed 28 November 2010).

48 Majors, R.E. (2007) Modern Techniques for the Extraction of Solid Materials – An Update, *LCGC Europe*, **20** (2).

49 Lapkin, A.A., Plucinski, P.K. and Cutler, M. (2006) *J. Nat. Prod.*, **69** (11), 1653–1664.

50 Naviglio, D., *et al.* (2007) *African J. Food Sci.*, **1**, 42–50.

51 Crown Iron Works Company, Desolventizer Toaster, Equipment Brochure, http://www.crowniron.com/userimages/Crown%20DTpdf2.pdf (accessed 28 November 2010).

7
Extraction of Lignocellulose and Algae for the Production of Bulk and Fine Chemicals

Thomas Hahn, Svenja Kelly, Kai Muffler, Nils Tippkötter and Roland Ulber

7.1
Introduction

Renewable resources will be an increasingly important issue for the chemical industry in the future. In the context of white biotechnology they represent the intersection point of agriculture and the chemical industry. The scarcity and related increase in the price of fossil resources make renewable resources an interesting alternative. If one considers the production of bulk chemicals it is evident that for this area besides the carbon sources, sugar and starch, new sources of raw materials must be accessible. One possible solution is to use lignocellulose and marine resources such as algae for the production of new materials and energy. Among the variety of microbial and chemical products that can be derived from glucose, lactic acid, citric acid, ethanol, acetic acid and levulinic acid are the preferred chemical intermediates for the synthesis of industrially relevant product families. For this both the development of new, as far as possible biologically degradable, products (e.g., lactic acid and levulinic acid reaction products), and also of intermediates in conventional product lines (acrylic acid, 2,3-pentandion) are viewed as possible strategies. Further important basic products are hydroxymethyl furfural (HMF) from the glucose line, furfural from hemicellulose and also substituted phenols from lignin. Among other things, furfural is a starting material for nylon. However, to be able to use renewable resources in the chemical, pharmaceutical or food industry, extraction processes are of tremendous importance. In this article methods for the extraction of lignocellulosic materials and algae to obtain bulk and fine chemicals are described. One focus lies on the use of new biotechnological procedures (biocatalysis and biotransformation), which enables simplified extraction, but well established methods are also explained. Lignocellulose and algae were chosen because of their growing importance both in academia and industry.

Industrial Scale Natural Products Extraction, First Edition. Edited by Hans-Jörg Bart, Stephan Pilz.
© 2011 Wiley-VCH Verlag GmbH & Co. KGaA. Published 2011 by Wiley-VCH Verlag GmbH & Co. KGaA.

7.2
Products from Lignocellulose

The biotechnological application of lignocellulose extracts for applications as biofuel production or replacement of petrochemicals is nowadays a high-priority goal of countries worldwide. Prior to their fermentation, the raw material polysaccharides have to be extracted and converted to sugars that are easily accessible for micro-organism growth and product formation. Lignocellulose biomass consists of three major fractions: cellulose with 35–50%, hemicellulose with 20–35% and lignin with 10–25% dry weight percentage. These main lignocellulosic compounds can be extracted via different procedures. Due to the high microbial and enzymatic degradation resistance of lignin, most extraction procedures for a subsequent biotransformation aim at the reduction of the lignin content.

Prior to the extraction processes mechanical treatment of the biomass is necessary to partly disrupt the structure and, more importantly, to increase the accessibility or the surface area of the substrate. The disruption is usually done by ball milling the wood. Desired substrate particle sizes mentioned in literature vary from millimeters to few centimeters. For very small particles up to one third of the power requirements of the entire biomass processing can be required for mechanical disruption of the raw material [1]. The principle processing scheme for a lignocellulose extraction and fractionation for subsequent biotransformations is shown in Figure 7.1.

7.2.1
Cellulose Fraction

Though polysaccharide hydrolysis can be done in acidic environments, enzyme-catalyzed synthesis of monosaccharides is favored. The application of enzymes has the advantages of very mild reaction conditions and lower maintenance costs for reactors compared with an acidic hydrolysis. Cellulase–complex mixtures extracted from *Trichorderma reseei* are capable of high glucose yields up to 75–85%. However,

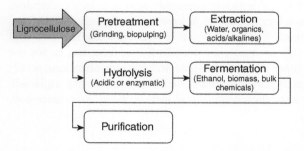

Figure 7.1 Processing scheme for the biotransformation of lignocellulose biomass.

the enzymatic reaction is considerably slower if the raw material has not undergone cleaning and fractionation steps.

From a biotechnological point of view the use of the glucose monomers of the cellulose polysaccharide is the straightforward approach for lignocellulose processing. A desirable function for successful and cost-competitive hydrolysis of the cellulose to its monomers would be high accessibility of the polysaccharide to catalysts – or more specifically, enzymes. Unlike the other primary renewable glucose sources, such as starch or sucrose, the cellulose polysaccharide is effectively protected against enzymatic attack by the surrounding matrix of lignin and hemicellulose.

In recent years several concepts have been proposed for lignocellulose extraction and subsequent processing of the derived sugars. Nevertheless, it seems reasonable to integrate and adapt existing wood processing methods from the pulp industry for pre-treatment of the lignocellulosic biomass. The separation of cellulose from the lignocellulose raw material is typically done by the ethanol organosolv process derived from the pulping process in the paper industry. Typical organosolv cellulose extractions are done at a temperature of 180 °C, with addition of 1.0–1.5% H_2SO_4 and ethanol contents in a range from 25 to 75% (v/v). Pan et al. [2] determined optimal organosolv extraction parameters of poplar chips for a subsequent optimal enzymatic hydrolysis of the cellulose obtained by response surface methodology (liquor to wood ratio 7 : 1, 180 °C, 60 min, 1.25% H_2SO_4, and 60% ethanol). After 60 minutes duration of the extraction procedure approximately 50% of the wood mass was obtained as solid with a cellulose recovery of 88%. 73% of the wood lignin could be dissolved. Most of the hemicellulose sugars (xylose, mannose, arabinose, and galactose) were found in the water-soluble fraction but less than half of the sugars were present in their monomer form. The enzymatic hydrolysis of the extracted cellulose was done with cellulase and β-glucosidase and resulted in a yield of 93% glucose after 24 h. In comparison, the enzymatic hydrolysis of softwood, pre-treated by steam explosion and still having a high lignin content of 42%, showed a glucose yield <30%.

While hemicellulose is relatively amorphous and can be readily enzymatically degraded by glycosidases, cellulose bound in lignocellulose occurs in a highly crystalline form. It is thereby more resistant against chemical and biological degradation. Therefore, a further aim of cellulose extraction, besides the reduction of the protective lignin, is its conversion into an amorphous form for easier enzymatic accessibility. Higher accessibility of the extracted crystalline cellulose can for example, be attained by treatment with an ionic liquid (IL). Several ILs are capable of dissolving cellulose and decreasing the degree of crystallinity after reconstitution [3]. Recently, the complete dissolution of lignocellulose materials (in this case Norway spruce sawdust) in ionic liquids as [Bmim]Cl and [Amim]Cl with subsequent enzymatic hydrolysis has been done [4]. The dissolution success is highly dependent on the size of the wood particles. Good results with ball-milled wood and sawdust were reported, whereas the dissolution of wood chips in ILs can require several weeks.

7.2.2
Hemicellulose Fraction

When separating the solid cellulose phase during a lignocellulose extraction process, a black liquor phase remains as residue. It is primarily composed of liquid lignin, hemicellulose and, depending on the specific extraction procedure, inorganic extraction chemicals. In conventional Kraft pulping the hemicellulose polysaccharides are degraded into mono- and oligosaccharides during the extraction process, then dissolve in the black liquor. In the classic pulping processing this liquor phase is used for steam and electricity generation.

As the hemicellulose has a considerably lower heat value than lignin, the use of the sugars for subsequent fermentation production seems reasonable. An overview of the primary products that can be biotechnologically derived from the lignocellulose biomass fractions can be found in Figure 7.2. Hemicellulose is a heterogeneous polymer consisting of several monomers such as pentoses (xylose, arabinose), hexoses (mannose, glucose, and galactose) and sugar acids [5]. Their fermentative application requires pre-extraction of the hemicellulose fraction or its separation from the lignin. Common extraction techniques are dilute acid pretreatment, alkaline extraction, hot water extraction and steam explosion or ammonia fiber explosion (AFEX).

At temperatures in a range from 200–250 °C, hot water can be used for hemicellulose extraction [6]. The hydrolysis and dissolution of the hemicellulose that occurs, is based on autohydrolysis. The reaction is catalyzed due to the formation of acetic acid by deacetylation of hemicellulose. The dissolved hemicellulose sugars can also contain minor amounts of already hydrolyzed glucose from the otherwise insoluble cellulose fraction. When the particle size of the substrate has

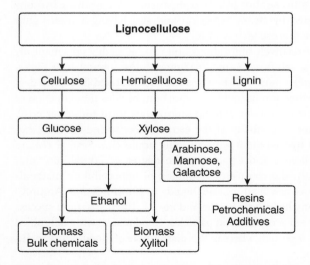

Figure 7.2 Biotechnological applications of the three main lignocellulose extracts.

been adequately reduced, xylose extraction rates of up to 90% can be attained by this method.

Application of water in form of a steam explosion by fast depressurizing of compressed steam can be used to reduce particle size of the substrates. Furthermore, hemicellulose and lignin are partly hydrolyzed or depolymerized [7], respectively, during the process. Approximately 50% of the substrate (primarily containing cellulose) remains as insoluble residue. Adding ammonia in the process of steam explosion results in the AFEX for lignocellulose extraction. It has the advantage of higher protein conservation and is especially useful for the hydrolysis of cellulose and hemicellulose in grass. Application to lignin-rich lignocellulose sources resulted in highly reduced yields [8].

By using dilute sulfuric acid in a concentration range from 0.5 to 1.0%, the temperatures of the aqueous hemicellulose extraction can be reduced to the range 150–200 °C for the extraction and liquefaction of most of the hemicellulose sugars [9]. Alkali extraction can be used for two-stage fractionations of lignin and hemicellulose. Lignin can be dissolved with 1% NaOH and 70% ethanol, removed and the residual hemicellulose extracted with 4% NaOH [10]. Combination of the different extraction procedures can increase the yield of the subsequent biotransformations. Nevertheless, the specific properties of the different lignocellulosic material sources must be kept in mind. Acid-catalyzed steam explosion, for example, is one of the most cost-effective processing methods for hardwood, but is much less effective when using softwood as substrate [11].

The isolated xylose monomers can be applied to several biotransformations, such as biomass and ethanol generation with micro-organisms, for example, *Pachysolen tannophilus*, *Candida shehatae* and *Candia tropicalis*. Furthermore, the sugar substitute xylitol can be fermented at high yields with xylose as substrate. Besides their use as a source of monomers for fermentation processes, hemicelluloses can be partly hydrolyzed to xylo-oligosaccharide (XO)s. These oligosaccharides are high value-added components in the field of functional foods, and have potential for example, as prebiotic additives. Furthermore, improvement of the sensory properties of foodstuff is possible with XOs [12]. While the extraction and preparation is possible with the hemicellulose processing methods described above, a more defined and specific production of the oligosaccharides is necessary using enzymatic processing. The extracted hemicelluloses can undergo controlled hydrolysis into oligosaccharides by addition of xyalanases with low exo-xylanase and/or β-xylosidase activity. For the subsequent refining of XOs in a defined molecular weight range, ultrafiltration has been described as a favorable method [13].

7.2.3
Extraction of Fermentation Inhibitors

Depending on the severity of the conditions during the extraction (including autohydrolysis), hemicellulose sugars can be degraded to aliphatic acids, furan derivates and phenolic compounds [11]. As degradation products of C_5 and C_6 sugars, primarily furfural and HMF are formed. For the subsequent fermentation procedures of the monosaccharides obtained, these inhibiting components should be removed from the extracts. Common methods for inhibitor removal are evaporation, solvent extraction, liming with calcium hydroxide, application of activated charcoal, ion exchange and enzymatic detoxification [14]. Application of enzymes is especially useful for the removal of phenolic compounds. This can be achieved by addition of laccases. Nevertheless, acetic acid, furfurals and HMF will still be present after the laccase treatment. Further evaporation or extraction procedures are needed to ensure complete inhibitor removal. Chandel *et al.* [15] examined ethanol yields during the fermentation of *Candida shehatae* NCIM 3501 with a lignocellulose extract containing inhibitors. Different inhibitor removal methods were compared. The highest ethanol yield with $Y_{P/S} = 0.48\,g\,g^{-1}$ was achieved after an ion exchange followed by inhibitor removal by activated charcoal ($0.42\,g\,g^{-1}$) and laccase ($0.37\,g\,g^{-1}$). Lowest yields were attained with overliming ($0.30\,g\,g^{-1}$) and neutralization ($0.22\,g\,g^{-1}$).

7.2.4
Lignin Fraction

Lignin is a highly branched, aromatic polymer, composed of phenylpropanoid units imparting rigidity, moisture and microbial resistance to lignocellulose. During enzymatic cellulose degradation, even already-defibrated lignin has a significant negative impact on the glucose yield, primarily caused by unspecific adsorption of the hydrolyzing enzyme on lignin particles. Lignin is often used as means of energy generation in form of thermal combustion during the processing of lignocellulose. Material recycling of the lignin for applications with industrial and consumer products, for example, as resins or concrete binders, seems cost-effective and, in the case of bioethanol production from lignocellulose, it can make the whole process competitive. Besides the previously described extraction methods for lignin, biological pre-treatment by white- and soft-rot fungi in the form of biopulping is possible. This depolymerization of lignin via micro-organisms is done at room temperature and does not require further addition of chemicals. It can be done in large wood chip piles or special solid-state fermenters (SSF). High moisture content of around 60% should be maintained during the biopulping. The mild reaction conditions make the process very energy and environmentally friendly. Nevertheless, this biological degradation takes a considerable amount of time and the fungi use some of the woods cellulose and hemicellulose for their own growth. Nevertheless, wood chips become softer and easier to disrupt after only one week of biopulping [16]. The biological processing seems cost-effective when handling high amounts of raw material and when used as a form of pre-treatment prior to further extraction procedures.

Despite the lignin disruption during biopulping, the fungi seem to initiate an esterification with oxalate produced by the organism, and thereby increasing the water saturation point of the cellulosic fibers. This weakens the fiber–fiber bonds and simplifies the subsequent attack of the lignocellulose polysaccharides [17].

When extracting lignin by the processes described previously, depolymerization and alteration of its structure often commences. If the lignin is intended for further use as an industrial raw material, milder reaction conditions should be chosen. Lee et al. [3] described the application of 1-ethyl-3-methylimidazolium acetate ([Emim][CH$_3$COO]) to maplewood for the very selective extraction of lignin from lignocellulose. Furthermore, the removal of 40% of the original lignin content by the ionic liquid was sufficient to reduce the crystallinity of the cellulose by approximately 30% and allowed cellulase hydrolysis of >90%. While all of the methods described above are based on liquid–solid extraction, complete solution of all wood components in a single step for subsequent liquid–liquid extraction is also possible. Lu and Ralph [18] describe the complete dissolution of wood in a mixture of dimethylsulfoxide (DMSO) and N-methylimidazole (NMI) at room temperature. In a subsequent step, a very pure lignin fraction can be precipitated from the solution.

7.2.5
Continuous Extraction of Lignocellulose

Kadam et al. [19] described a continuous fractionation counter-current process using a twin extruder. In the first stage, lignocellulose can be fed into the systems via a high pressure pump. Subsequent autohydrolysis of the substrate is done at a temperature of 210 °C. 60–65% of the hemicellulose sugars which are extracted, enter a counter-current liquid flow and exit the system. The solids enter a second extruder and undergoing an alkali extraction. The resulting lignin liquor and cellulose solid are discharged from the system via a liquid–solid separation. The final solid fraction has a glucan proportion of 85–90%.

7.3
Polysaccharides and Sporopollenin from Marine Algae

Red and brown seaweeds produce hydrocolloids like agar, alginate and carrageenan, which are used for example, in the food and cosmetic industry. Approximately 55 kt of these polysaccharides are extracted from 1000 kt of wet seaweed each year (value: US$ 600 million) [20] to yield in about 23 kt alginate, 7.5 kt agar and 28 kt of carrageenan [21]. This chapter gives an overview of some algal polysaccharides of industrial interest, their sources, respective extraction methods and applications.

7.3.1
Agar

The designation "agar" is descended from the Malayan word "agar-agar" and means jelly. Agar is a hydrocolloidal compound primarily found in marine red algaes (*Rhodophyta*) especially in the species of *Gelidium*, *Gracilaria* and

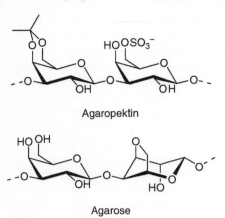

Figure 7.3 The two main compounds of agar; here: agaropectin with a sulfuric acid residue at 6-position and a pyruvic acid ketal substituent.

Pterocladia, sometimes called agarophytes. The amorphous cell wall components including agar and other galactans surround fibrils consisting predominantly of cellulose [21, 22]. Agar is a complex polysaccharide, which consists of up to 70% agarose and 30% agaropectin. It has a molecular weight of about 110 000–160 000 g mol^{-1} depending on the origin (type of seaweed). The main component agarose is an unbranched polymer consisting of the structural units D-galactose and 3,6-anhydro-L-galactose, which are alternately attached in an α-1,3 and β-1,4 linkage. Agaropectin is a slightly branched more complex polymer of β-1,3 glycosidic linked D-galactose partly sulfated at position 6 (3% to 10% sulfate content). It contains methyl and pyruvic acid ketal substituents in acetyl linkages as well as glucuronic acid conjugated to agarobiose (Figure 7.3) [22, 23]. The component with the higher swelling tendency is agarose, thus it is desirable to separate the polymers in fractions. The separation of agarose and agaropectin can be based on the difference in solubility of the acetal derivates.

In the food industry agar (food additive E 406) is used as thickening agent and can be applied as substitute for pectin, gelatin and starch. The best known application of agar is the preparation of culture media in Petri dishes for the growth of micro-organisms, although only 20% is used for biotechnological applications [24]. Agars form a crucial carbon source for several marine micro-organisms. To digest the agars, the micro-organisms need to secrete extracellular enzymes, like agarases (α-agarase EC 3.2.1.158; β-agarase EC 3.2.1.81), which hydrolyze the polymers into oligosaccharides. Moreover these biocatalysts are tools to design new biomolecules with various length and molecular weight and therefore altered properties, to produce new products for the food, cosmetic and pharmaceutical industries [25].

An application of agars, for example in functional food, requires a highly purified product. The first step in this procedure is the extraction of the natural compound. A variety of methods optimized for small-scale agar extraction are described for the red algae [25–28] in literature, right up to ecofriendly [29, 30] and various

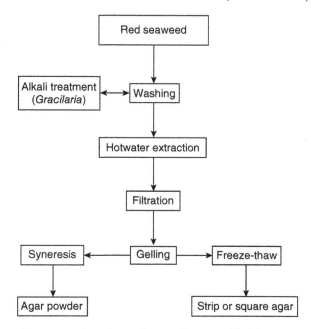

Figure 7.4 Industrial manufacture of agar (modified from [36]).

pre-treatment procedures [31, 32]. Industrial-scale extraction depends on the method applied and the algal source, too. On the other hand the main steps in extraction and purification are comparable. The first step in extraction is mechanical cleaning of the algae followed by washing to remove foreign matter. Before hot water extraction is performed, algae of the species *Gracilaria* have to be pre-treated with alkaline solution, to cause a chemical change in the agar structure. This provides a higher gel strength for commercial use based on desulfation, which results in formation of 3,6-anhydrogalactose bridges [33]. To dissolve the agar, extraction is carried out in boiling water usually under pressure. pH adjustment can be performed with dilute sulfuric acid or acetic acid to promote better extraction [24]. Afterwards, impurities and other algae raw material are removed from the liquid extract by the application of standard industrial filtration units. During extraction it is important to work at raised temperatures preventing the liquid from gelling. After extraction the liquid cools down, building a gel containing about 1% of agar [34]. The gel can be bleached to decrease the color, the bleaching agent is removed as well as salts by osmosis with washing water [35]. There are two different methods to remove the water from the agar: the freeze–thawing method and the press-dehydration or syneresis method (Figure 7.4). The fundamental principles of the former technique were first recorded by Tarazaemon Minoya in 1658 and rely on the insolubility of cool agar [35]. Slow natural or mechanical freezing results in the formation of large ice crystals, whereby the gel structure is broken down. During thawing, the water with residual salts, smaller carbohydrates and proteins and pigments drains off [36]. The process results in a tenfold increase of the agar concentration and gives strip or square agar.

The other process is called the syneresis or press-dehydration method. It is based on the property of gels to release water by application of pressure. The 1% agar gel is placed between two metal plates covered with cloths. An increasing force is applied to the metal plates to put the gel under pressure. Nowadays, the method is totally automated and gives a gel with agar concentration of about 20%. Drying and milling results in agar powder [36]. The syneresis method demands higher investment costs but has lower energy consumption during the process, in contrast to the freeze–thaw method.

7.3.2
Carrageenan

It is generally accepted that the name carrageenan has an Irish origin. One thesis implies that the root of the word comes from an Irish coastal village, the other claims that the word is based on the Irish root word "carraigeen" (engl. Moss of the rock) [37, 38]. Carrageenan is a collective term for linear sulfated hydrocolloids obtained from the cell wall of red seaweeds like *Chondrus crispus* (Irish Moss) and *Mastocarpus stellatus*. For carrageenan production in the Philippines, species of the genus *Eucheuma* are cultivated [30]. Like the amorphous fraction of the cell wall of agarophytes consisting of agar, the amorphous fraction of the carrageenophytes' cell wall is mainly composed of carrageenans. The term carrageenan includes several salts of linear heterogeneous galactans. The monomer building blocks consist of galactose units which are partially sulfated and/or are present in the anhydrogalactose form. The algal polysaccharides have molecular weights from $100\,000$–$1\,000\,000\,g\,mol^{-1}$, whereas molecular weights, structures and chemical composition depend on the species of the algae and their life-stage [39, 40]. The commercially important polymers are κ (kappa)-, ι (iota)- and λ (lambda)-carrageenans; the β-, γ-, μ-, ν-, ξ- und θ-forms also exist [41]. The main repeating dimeric units are (1→3)-β-D-galactopyranose-2-sulfate-(1→4)-α-D-galactopyranose-2,6-disulfate in λ-carrageenan and (1→3)-β-D-galactopyranose-4-sulfate-(1→4)-3,6-anhydro-α-D-galactopyranose in κ-carrageenan, whereas ι-carrageenan is the 2-sulfated form (Figure 7.5).

The κ- and ι-carrageenans build gels with cations in a brittle or elastic form and in different colors depending on the counter-ions, and ι-carrageenans usually aggregate in softer and more elastic gels than κ-carrageenans do. λ-carrageenans do not form gels with cations because of the lack of the 3,6-anhydro linkage, which prevents the initial double helix formation that is responsible for gel formation. Moreover, the number and position of the ester sulfate groups on the repeating galactose units primarily determine the gelling properties of the carrageenans [42, 43].

Carrageenans are common food additives (E 407) due to their thickening, gelling and emulsion stabilizing properties in the food industry [44]. κ-carrageenans are the most used sulfated galactans in a wide range of products and can be used

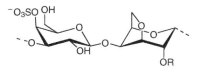

Figure 7.5 The dimeric repeating units of the three most important carrageenans.

R= SO$_3^-$: ι–Carrageenan; R=H : κ–Carrageenan

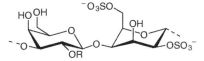

R= 70 % SO$_3^-$, 30 % H : λ–Carrageenan

solely or in synergy with locust bean gum or konjac flour as well as ι-carrageenans with starch [45]. Furthermore, they can be applied in the pharmaceutical or cosmetic industry. In order to use the algal carrageenan as a crucial carbon source, bacteria of the classes *Gammaproteobacteria*, *Flavobacteria*, or *Sphingobacteria* that release specific glycosidase hydrolases, the so called carrageenases (κ-carrageenase EC 3.2.1.83; ι-carrageenase EC 3.2.1.157; λ-carrageenase EC 3.2.1.162) are used. The enzymes known so far act specifically on the β-(1→4)-linkage of the substrates [46].

An extraction and purification procedure was firstly patented by Baird in 1871 [47]. Thenceforward, improvement of the known extraction conditions with various algae on the laboratory scale was the object of research in order to get higher yields and high quality products. Nowadays, scientists turn the spotlight more and more on alternative and ecofriendly extraction methods [48, 49]. Two main industrial procedures are used to obtain carrageenan. The first, the only one used until the 1980s, is based on the dissociation of carrageenans into an aqueous solution and results in refined carrageenan. The second method rests on the extraction of unwanted algal substances into a potassium hydroxide solution (Figure 7.6) Carrageenan and cellulose comprise the insoluble residue, together called semi-refined carrageenan (SRC) or alkali-modified flour (AMF). The original method starts with extraction of carrageenans from washed algae with a hot alkaline solution and can take up to a day [38]. As with the alkali pre-treatment in agar extraction, the base modifies the chemical integrity and gives a high quality product. Two filtration steps and/or high speed-centrifugation were performed to remove raw algal material and to clarify the extract [37]. Concentration from 1–2% to 2–3% carrageenan can be performed with common methods like distillation [42]. To recover the product, precipitation with isopropyl alcohol is carried out, contaminants remain in solution. Drying and milling results in pure carrageenan. For κ-carrageenan in particular, the facility is provided to form a gel in the presence of potassium chloride. The residual water can be removed either with press-dehydration or freeze–thawing. The SRC or AMF is the more economic and efficient procedure with respect to the classical extraction [47]. The freshly harvested

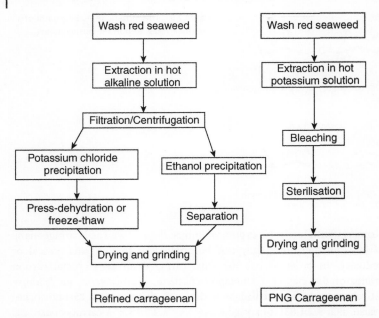

Figure 7.6 Industrial manufacture of carrageenans; left side: chart for the production of refined carrageenan; right side: chart of the production of PNG carrageenan (modified from [47]).

algae are directly treated with a potassium hydroxide solution, and the temperature of the solution does not exceed 70–80 °C in order not to dissolve the carrageenans. The hydroxide causes splitting of the sulfate groups and therefore the building of 3,6-anhydro bridges in the algal carrageenans, while the potassium forces gel formation and thus insolubility. After washing and drying the gel is chopped and milled to the desired size. The process is mainly used for κ-carrageenan because it forms rigid gels with potassium. If the SRC obtained is to be used as a food ingredient, post-processing is necessary. The two further steps are bleaching to reduce the color and sterilization with hot steam [50]. The resulting carrageenan–cellulose mixture is sold as Processed *Euchema* Seaweed (PES) or Philippines Natural Grade (PNG).

7.3.3
Alginic Acid

Alginic acid is a structural polyuronide of the cell walls of brown seaweed (*Phaeophyceae*), which is responsible for the high mechanical flexibility of the algae. It may constitute 15 to 40% of the dry weight of these algae. The salts of alginic acid are called alginates, which easily undergo cation exchange reactions. Depending on the type of associated ions the properties of the alginate may vary considerably. Salts of the alkali metal ions (e.g., sodium alginate) are water soluble, whereas alginic acid itself and its salts with trivalent and divalent metal ions (e.g., calcium alginate) are generally insoluble and able to form gel matrices. The possibility of influencing the structure and behavior of these gels by variation of the types of associated ions is the reason for diverse applications of alginates in several fields of industry. The most important algae species for commercial alginate production are *Macrocystis pyrifera* (giant kelp, American Pacific coast), *Ascophyllum nodosum* (American Atlantic coast and Europe, e.g., Scotland, Ireland, Norway), *Laminaria* species (*L. hyperborean* and *L. digitata*: Europe; *L. japonica*: Japan) and *Ecklonia* species (South Africa, South Australia) [51, 52]. Furthermore two bacterial genera are known to have the possibility to synthesize alginates, *Pseudomonas* and *Azotobacter* [53]. The cultivation (farming) of brown seaweed is expensive and time-consuming but profitable for the application of the entire algae in the food industry. On the other hand, algae for industrial alginate production are harvested from naturally growing seaweed because cultivation would not be economically viable due to the additional downstream processes required for the alginate extraction. Specifically designed vessels equipped with rotating cutters or curved iron hooks, rake-like devices, suction pipes or sloping wire mesh belts are used to harvest the algae mechanically. Algae growing in low tide areas are still cut by hand as well [52].

Alginate is a linear polysaccharide consisting of two types of anionic monomers. β-D-mannuronic acid (M) and α-L-guluronic acid (G) are linked via 1,4-glycosidic bonds in homopolymeric MM- and GG-blocks respectively as well as regions of alternating monomers (MG-/GM-blocks) as indicated in Figure 7.7. Alginates form gels at low pH values or in the presence of, for example divalent cations, such as Ca^{2+}, which link specifically to the G blocks by binding to the free carboxyl groups and giving rise to a three-dimensional network. The higher the ratio of GG-blocks in the alginate, the higher is the stability of the resulting gel. On the other hand, the alternating blocks form the most flexible chains, and alginates rich in MG-blocks are less viscous and exhibit a higher solubility in acids [54, 55]. While the viscosity of alginates depends on their molecular weight, the gel-forming and water-binding properties depend on the relative proportion of the three types of blocks, which vary with the seaweed source and seasonal variations [56].

Alginate (E400–E404) is widely applied as a stabilizing, thickening or emulsifying agent in the food, cosmetic, paper and dye industries. In medical fields alginates are used for example as dental impression materials [57], for tissue engineering applications [58] or as cell encapsulation material [59]. Alginates are also applied as ingredients in absorbent dressings. Due to their ion exchange reactions and their water-binding capacity alginates establish ideal conditions for wound-healing, especially for exuding wounds [60]. Furthermore, an incorporation of silver ions into alginate fibers leads to wound dressings with antimicrobial properties [61]. As it is indigestible, alginate is studied as dietary fiber as well and may protect against cardiovascular and gastrointestinal diseases [57]. It is also used in anti-reflux therapies as remedy for heartburn and oesophagitis, as it forms a floating gel in contact with gastric acid thus providing a relatively physical, pH-neutral barrier [62]. Alginate-degrading enzymes are mannuronate (EC 4.2.2.3) or

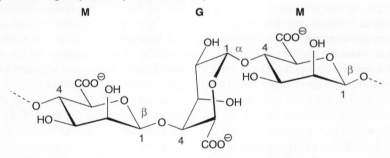

Figure 7.7 Structural block types of alginates (modified according to [25]).

guluronate lyases (EC 4.2.2.11), which are produced by many bacteria, for example, *Sphingomonas sp.*, *Alteromonas sp.* and *Corynebacterium sp.* [63].

The definition of separation and purification steps throughout the extraction process is facilitated by the ability of alginate to vary between aqueous phase and gelation via cation exchange reactions. The harvested algae were traditionally washed, dried and milled. As drying is very expensive, processing methods with wet algae were increasingly applied. In this case, the washing of the raw material is followed by stabilization with formalin, mineral acids and preservatives [64]. Treatment with formalin additionally causes a binding of the otherwise water-soluble polyphenolic compounds of the algae, which remain insoluble and may be separated from soluble alginates later on, resulting in less-colored alginates [55, 65]. The alginate matrix of the algal cell wall is composed of insoluble calcium, magnesium and sodium alginates. Converting these alginates into soluble sodium alginates finally enables extraction from seaweed residues. For this purpose, the formalin pre-treated algae often undergo an acid pre-extraction treatment, which first converts the alginates of the algae into insoluble alginic acid. The conversion of alginic acid into sodium alginate is much more efficient than the conversion of the original calcium alginate [65]. Soluble algae components like laminaran, fucoidan, oligo- and monosaccharides and amino acids are removed along with the acid treatment [64]. The remaining solid seaweed residues consist of alginates, pigments, cellulose and proteins. Extraction with a hot sodium carbonate solution converts the alginates into soluble sodium alginate. The liquid phase containing alginates and pigments can be separated from proteins and cellulose by filtration, centrifugation or flotation steps [54, 64]. But as an acid pre-extraction may decrease the viscosity of the resulting alginates, processes are also proposed in which the alginates are directly extracted with sodium carbonate under strong alkaline conditions (pH 10, 80 °C, 2 h) [65]. However, the high viscosity of the resulting sodium alginate extract hampers the separation from seaweed residues after extraction and requires a high dilution of the viscous slurry to enable mechanical separation processes like filtration, centrifugation or flotation steps. A subsequent precipitation with acid or calcium chloride converts the highly diluted sodium alginates into insoluble alginic acid and calcium alginate respectively and regains the alginate in a considerable smaller volume [66]. The pigments do not precipitate and remaining traces in the precipitate can be removed by washing with acid and calcium chloride respectively. Figure 7.8 gives an overview of the extraction process. According to industrial demand the purified and concentrated alginates are converted with the respective carbonates to sodium, potassium, ammonium, magnesium or calcium alginate and are dried and milled.

7.3.4
Fucoidan

Fucoidan, also a polysaccharide of brown seaweed, is of increasing pharmaceutical interest although it is not yet produced industrial. Fucoidans are found in the cell walls of all *Phaeophyceae*, but the amount varies seasonally and depends primarily on the algae species and the depth at which they occur. Algae of low tide areas (e.g., *Fucus vesiculosus, Pelvetia canaliculata*) exhibit more fucoidan than permanently submerged species (e.g., *Laminaria* sp.). According to that, fucoidan seems to protect the algae from dehydration. It is highly hygroscopic and forms, along with soluble alginates, the amorphous matrix of the cell wall, whereas the fibrillar fraction of the cell wall consists of cellulose and insoluble alginates.

Fucoidans are sulfated polysaccharides mainly consisting of L-fucose with molecular weights ranging from 13 to 950 kDa [65]. Like their molecular weights, their complex and heterogeneous structures vary considerably from algae to algae

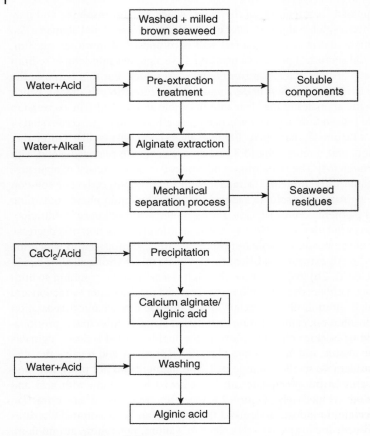

Figure 7.8 Production of alginic acid.

and therefore complicate overall structure determinations and characterizations. Often proposed is a branched backbone of α-(1→3)-linked or α-(1→4)-linked L-fucopyranosyl residues which may be sulfated mainly at position 2 or 4 [66–69]. Further substituents are acetyl groups or monosaccharides like galactose, glucose, mannose, xylose, or glucuronic acid. Figure 7.9 gives an example of a α-(1→3)-linked fucoidan.

Fucoidans are of pharmaceutical interest due to their versatile bioactivity. For example, they exhibit anticoagulant abilities by enhancing the heparin cofactor II, so they may become an alternative to heparin due to their herbal origin. They are also widely studied because of their antitumor efficacy, owing to their ability to inhibit proliferation and to induce apoptosis via caspase and extra-cellular signal-regulated kinase pathways [20, 30]. Their ability to hamper the replication of viruses, or their penetration into host cells results in antiviral activities and yields positive effects for example, against human immunodeficiency virus (HIV) and herpes simplex virus type 1 (HSV-1) [70]. It has been shown that enzymatically-

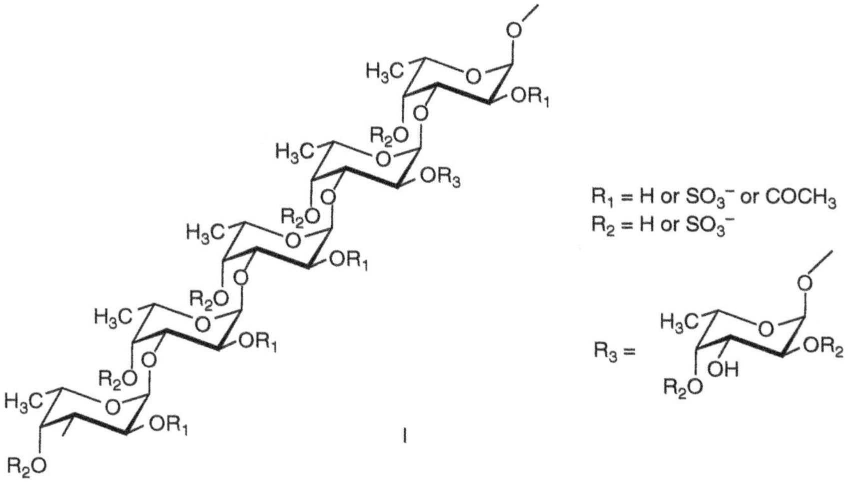

Figure 7.9 Structure of fucoidan, (1→3)-L-fucopyranosyl (according to [69], supplementary data).

R_1 = H or SO_3^- or $COCH_3$
R_2 = H or SO_3^-

hydrolyzed fucoidans exhibit enhanced bioactivity. Enzymes capable of hydrolyzing the glycosidic bonds of fucoidans without release of sulfate were found in marine invertebrates and bacteria and are called fucoidanases (EC 3.2.1.44, [71]) and α-fucosidases (EC 3.2.1.51, [72]). Whereas α-fucosidases release L-fucose from the non-reducing end of fucoidans, fucoidanases can exhibit endo-hydrolyzing and exo-hydrolyzing activities respectively. Further investigations are being undertaken to elucidate the correlation between structure and bioactivity of fucoidans, to enable the production of fucoidans with high pharmaceutical abilities via directed enzymatic hydrolysis.

Different extraction methods are described in literature [73–76], but mostly the first steps of the fucoidan extraction are almost analogical to the alginate extraction concerning for example, a pre-treatment with formalin and an extraction under acidic conditions. The seaweed is for example, pre-treated overnight with an ethanol/formaldehyde/H_2O solution (80:5:15 v/v) and the dried algae is subsequently extracted twice with a 0.01 N HCl solution supplemented with 4% $CaCl_2$ (w/v) for 3 h at 70 °C, leaving alginates, proteins and cellulose insoluble [72]. As mentioned above, laminaran, oligosaccharides, monosaccharides and amino acids are extracted as well. A treatment of the extract with 2.5 volumes of ethanol precipitates the polysaccharides, so that they are separated from the smaller components. As laminaran is a polysaccharide which does not naturally contain sulfate groups [77], fucoidan and laminaran may be separated by ion exchange chromatography and fractionated via gel filtration chromatography.

The many applications of algal polysaccharides in industry emphasize their importance as seen in the worldwide annual output figures. Especially in food industry the polysaccharides take on greater significance which relies mainly on their properties to alter the consistency of food in conjunction with beneficial health effects as requested in "functional food". Actually, in the near future, extraction of the polysaccharides from algae will be the method of choice, to produce these natural compounds because of their chemical diversity, which impedes the classical organic synthesis of these products. Improvement of the previously mentioned extraction methods is not yet finished, focusing on ecofriendly and economical developments or completely new extraction methods, for example the production of semi-refined carrageenan.

7.3.5
Sporopollenin/Algaenan

The extraction of algal strains delivers outstanding high valuable compounds as mentioned before. However, such ubiquitous algal material can be also used to obtain specific components primarily found in terrestrial plants, which could be commercialized as nutraceuticals, pharmaceuticals and/or additives for high end cosmetics, respectively. As an example the cell wall component sporopollenin (also termed as algaenan) can fulfill the respective requirements. Therefore its extraction from biomass is considered in more detail in this section. The use of aquatic biomass can be very beneficial, because the lignin fraction of these plants is commonly lower than those of terrestrial origin. Additionally, the exploitation of renewable raw materials such as algae biomass, is currently not the topic of controversy debates about nutrition conflicts, which are fervidly considering the substitution of gasoline by the first generation of biofuels.

Sporopollenin is a specific plant (hetero-)biopolymer which was originally found in the outer walls of spores and pollen grains. The name was originally introduced by Zetsche and Viceri in 1931, according to their investigations of spores and pollens of *Lycopodium clavatum* [78]. It consists of different fractions of long chain aliphatic and aromatic sub-units, where the ratio of both depends on the origin of the biogenous source. Based on ^{1}H-NMR analysis of sporopollenin extracted from pollen of Narrow Leaf Cattail (*Typha angustifolia*) Ahlers *et al.* [79] proposed different fundamental phenolic structures which are presented in Figure 7.10.

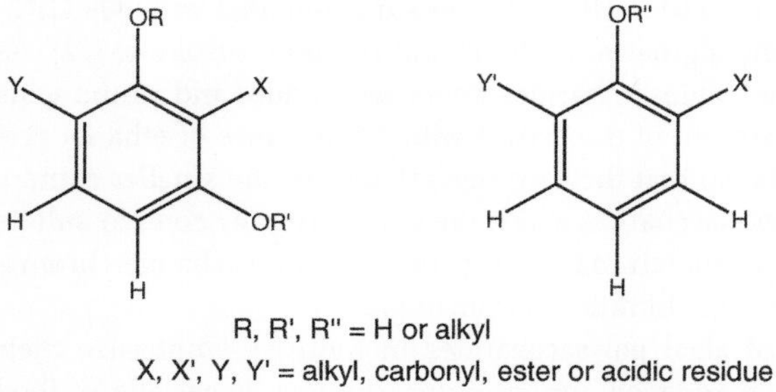

R, R', R" = H or alkyl
X, X', Y, Y' = alkyl, carbonyl, ester or acidic residue

Figure 7.10 Proposed structure of sporopollenin from Narrow Leaf Cattail origin [80].

Due to its specific structure sporopollenin is highly resistant to chemical degradation and furthermore to solubilization. The latter can be effectively carried out by using aminoethanol. The complete biological function has not yet been elucidated, but it is assumed that the biopolymer plays a key role in protecting the pollen and other plant materials from mutagenous UV-radiation [80]. Furthermore, the compound was found to efficiently support detoxification processes within the human metabolism [81, 82], especially of heavy metals and halogenated organic compounds. However, the occurrence of sporopollenin is not limited to spores or plant pollen and can be also found in other parts of plant material such as algal cell walls. Therefore several submers growing algae can be considered as a renewable resource for the extraction of sporopollenin. A promising approach could be based on the exploitation of extracts of the ubiquitous algae *Chlorella vulgaris*, because this strain can be cultivated in a larger scale in process plants (open ponds, raceways, closed bioreactor concepts), while the use of a technical bioreactor provides the opportunity to achieve a monoseptic cultivation of the algae. Such procedures are necessary, because naturally occurring biomass material is often polluted and does not show the required purity for pharmaceutical or technical purposes. A phototrophic as well as mixotrophic growth of *C. vulgaris* is possible. If the optimal conditions for growth are adjusted, the risk of contamination can be minimized. The high content of sporopollenin of *C. vulgaris*, which amounts up to 3–8% of the biomass is significant. If one has a closer look at the literature several publications named *Chlorella pyrenoidosa* as a good source for sporopollenin extraction, whereas *C. vulgaris* is assumed to contain only a minor fraction of the biopolymer. To avoid any confusion about the distinct names it is worth mentioning that the formerly strain *C. pyrenoidosa* has been newly classified as *C. vulgaris* [83, 84].

In principle, the sporopollenin has to be efficiently released by the cell walls of the organic matter within the isolation process. Prior to the application of chemical or enzymatic methods, necessary for degradation of the cell walls, it is therefore reasonable to use mechanical methods to grind (e.g., ball mill) the dried biomass and provide a bigger surface for the subsequent extraction methods and to raise the extraction efficacy, respectively. However, due to its chemical resistance, the common extraction approach of sporopollenin from biomass material starts with treatment of the organic matter by a defined mixture of sulfuric acid and acetic anhydride (in a ratio of 9:1), a process which is named acetolysis [85, 86]. Alternatively, the enzymatic cell wall degradation process can be carried out under much less harmful conditions [87]. But the applied approach has to be adapted to the special needs of the whole process. Whereas more or less most of all of the interior components are oxidized within the acetolysis, the cellulolytic degradation of the cell provides the release of side products in their native states within the process. Thus, such by-products can also be obtained and therefore represent additional economical target compounds, such as β-carotene or chlorophyll. If one has a closer look at the content of both natural compounds, which amounts up to 2–3% for chlorophyll and several percent for β-carotene, beneficial economical side-effects are obvious; the current price is approx. 117 € mg^{-1} for chlorophyll from

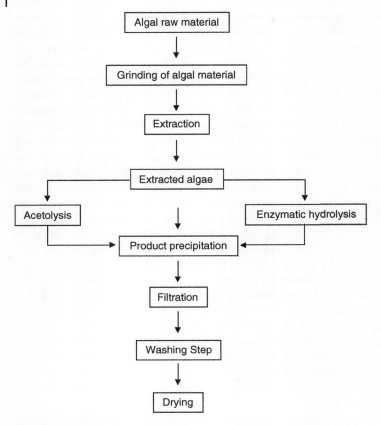

Figure 7.11 Flowsheet of sporopollenin recovery from algal material.

the algae *Anacystis nidulans* and approx. 63 € for mixed isomers of carotene from a natural resource (carrots). But even the oxidized side-products hamper the isolation of the sporopollenin, due to the additional downstream steps that are required. Within a sophisticated purification approach it is possible to get rid of the chlorophyll content by extraction of the biomass matter with organic solvents such as dimethyl sulfoxide, acetone, isopropanol or ethanol, respectively. A flow sheet is presented as an example in Figure 7.11.

As described in Figure 7.11 the algal raw material has to be ground and the chlorophyll content should be separated from the biomass. The milling process could be efficiently carried out by using a ball mill for 4 h with a frequency of mechanical oscillation of 1800 min^{-1}, while the subsequent chlorophyll extraction is achieved by treatment of the ground algal biomass with dimethyl sulfoxide (1 g of biomass per 300 mL DMSO) at 60 °C. The extracted algal matter can subsequently be applied for the acetolysis process. This process can be carried out within the frame of an optimized procedure. Thus, a well defined mixture of acetic

Figure 7.12 SEM micrograph of the *C. vulgaris* algal raw material (a) and fractions obtained during the recovery process of sporopollenin; (b) treated with DMSO; (c) after acetolysis.

anhydride and sulfuric acid of 1:10 (v/v) is prepared, which is afterwards used for treatment of the biomass within a ratio of 10:1 (v/w) for 2 h at 60 °C and the sporopollenin content is efficiently released from the cell walls. The resulting dark-colored solution is diluted with algid water by approximately 1:10 (v/v), when the target compound precipitates and can easily be separated by filtration. A further washing step and drying results in a pale yellow solid matter with a yield of about 8% concerning the biomass input. Quality control of the degradation was done by application of electron microscopy to the input-fraction of *C. vulgaris* and the obtained fractions after treatment of the biomass with the organic solvent and the acetolysis mixture, respectively. The size and shape of the resulting matter is shown in Figure 7.12. Furthermore, an UV/Vis-spectrophotometric analysis of the extract indicates that the sporopollenin fraction was effectively removed by organic phase extraction (see Figure 7.13).

As already described in Figure 7.11 the algal raw material has to be treated in the same manner as the preparation for the acetolysis process, that is, the algal raw matter has to be ground and the chlorophyll content is separated from the biomass under the same conditions. A sustainable enzymatic hydrolysis procedure can be applied by treatment of the ground and extracted biomass material with

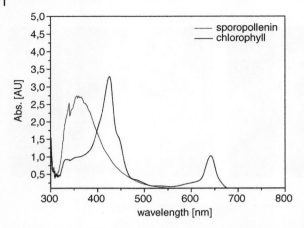

Figure 7.13 Spectra of acetolysis product and chlorophyll (both compounds were solubilized in 2-aminoethanol).

technical enzymes such as Alcalase® and Flavouzyme® (both from Novozymes, DK). The hydrolysis can be carried out under slightly basic conditions, preferably at pH 8.5 for four days at 31 °C. Subsequently, after this enzymatic cell wall degradation method, the insoluble sporopollenin fraction can be separated by filtration or centrifugation, whereas the recovered solid matter is treated with 4% KOH (w/v) and 0.2% $Na_2S_2O_4$ (w/v) for 4 h followed by washing with 0.1% sodium dodecyl sulfate and three times with water. Removal of remaining lipids was achieved by a further washing step with acetone, isopropanol and water, whereas a sporopollenin-fraction of 8% results related to the algal raw material used. However, the recovered product fraction shows the same pale yellow color and approximately the same yield compared with the fraction which is obtained from the chemical hydrolysis process. The enzymatic hydrolysis approach shows a couple of considerable advantages in comparison with the acetolysis procedure. This can be summarized as follows: application of very mild conditions, use of considerably less harmful reaction compounds, recovering procedure generates minimized toxic waste streams and does not need complex and expensive neutralization/detoxification units.

References

1 Wooley, R., et al. (1999) Lignocellulosic biomass to ethanol – Process design and economics utilizing co-current dilute acid prehydrolysis and enzymatic hyrolysis – Current and futuristic scenarios. Report No. TP-580-26157. National Renewable Energy Laboratory, Golden Colorado USA, 130.

2 Pan, X., et al. (2006) Biotechnol. Bioeng., **94** (5), 851.

3 Lee, S.H., et al. (2009) Biotechnol. Bioeng., **102** (5), 1368.

4 Kilpelainen, I., *et al.* (2007) *J. Agric. Food. Chem.*, **55** (22), 9142.
5 Saha, B.C. (2003) *J. Ind. Microbiol. Biotechnol.*, **30** (5), 279.
6 Mok, W.S.L. and Antal, M.J.J. (1992) *Ind. Eng. Chem. Res.*, **31** (4), 1157.
7 Cara, C., *et al.* (2006) *Process Biochem.*, **41** (2), 423.
8 Sun, Y. and Cheng, J. (2002) *Bioresour. Technol.*, **83** (1), 1.
9 Knappert, D.R., Grethlein, H.E. and Converse, A.O. (1981) *Biotechnol. Bioeng.*, **11**, 67.
10 Gabrielii, I., *et al.* (2000) *Carbohydr. Polym.*, **43** (4), 367.
11 Hamelinck, C.N., van Hooijdonk, G. and Faaij, A.P.C. (2005) *Biomass Bioenergy*, **28** (4), 384.
12 Voragen, A.G.J. (1998) *Trends Food Sci. Technol.*, **9** (8–9), 328.
13 Akpinar, O., *et al.* (2007) *J. Agric. Food Chem.*, **55** (14), 5544.
14 Huanga, H.-J., Ramaswamy, S., Tschirner, U.W. and Ramarao, B.V. (2008) *Sep. Pur. Techn.*, **62** (1), 1.
15 Chandel, A.K., Kapoor, R.K., Singh, A. and Kuhad, R.C. (2007) *Bioresour. Technol.*, **98** (10), 1947.
16 Ferraz, A., *et al.* (2008) *Enzyme Microb. Technol.*, **43** (2), 178.
17 Hunt, C., Kenealy, W., Horn, E. and Houtman, C. (2004) *Holzforschung*, **58** (4), 434.
18 Lu, F. and Ralph, J. (2003) *Plant J.*, **35** (4), 535.
19 Kadam, K.L., Chin, C.Y. and Brown, L.W. (2008) *J. Ind. Microbiol. Biotechnol.*, **35** (5), 331.
20 Hoek, C.V.D., Mann, D.G. and Jahns, H.M. (1995) *Algae: An Introduction to Phycology*, Cambridge University Press, Cambridge, UK.
21 Vreeland, V. and Kloareg, B. (2000) *J. Phycol.*, **36** (5), 793.
22 Steglich, W., Fugmann, B. and Lang-Fugmann, S. (1997) *Römpp Encyclopedia of Natural Products*, (eds J. Falbe and M. Regitz), Thieme Verlag, Stuttgart, Germany, p. 9.
23 Nussinovitch, A. (1997) Agar, in *Hydrocolloid Applications: Gum Technology in the Food and Other Industries*, Black Academic & Professional, London, UK, Ch 1, pp. 1–18.
24 Arvizu-Higuera, D.L., *et al.* (2008) *J. Appl. Phycol.*, **20** (5), 515.
25 Barbeyron, T., *et al.* (2006) *Appl. Microbiol. Biotechnol.*, **71** (1), 23.
26 Pareira-Pacheco, F., Robledo, D., Rodriguez-Carvajal, L. and Freile-Pelegrin, Y. (2007) *Bioresour. Technol.*, **98**, 1278.
27 Chiovitti, A., *et al.* (2004) *J. Appl. Phycol.*, **16** (1), 41.
28 Villanueva, R.D., Pagba, C.V. and Montano, N.E. (1997) *Botanica Marina*, **40** (5), 369.
29 Li, H., *et al.* (2008) *Bioresour. Technol.*, **99** (8), 3301.
30 Li, H., *et al.* (2009) *J. Appl. Phycol.*, **21** (2), 247.
31 Roleda, M.Y., *et al.* (1997) *Botanica Marina*, **40** (1), 63.
32 Murano, E., *et al.* (1993) *J. Appl. Phycol.*, **5** (4), 417.
33 FreilePelegrin, Y. and Robledo, D. (1997) *Botanica Marina*, **40** (4), 285.
34 Armisen, R. and Galatas, F. (2000) Agar, in *Handbook of Hydrocolloids* (eds G.O. Phillips and P.A. Williams), Woodhead Publishing Limited, Cambridge, UK, Ch 2, pp. 21–40.
35 McHugh, D.J. (2003) Agar, Chapter 3 in Fisheries Technical Papers 441: A Guide to Seaweed Industry, FAO, Rome, Italy, pp. 17–26.
36 Coppen, J.J.W. and Nambiar, P. (1991) *Agar and Alginate production from the seaweed in India*, in Post-Harvest Fisheries, Food and Agriculture Organisation of the United Nations, FAO, Madras, India.
37 Voragen, A.C.J. and Pilnik, W. (1994) Other polysaccharides, in *Ullmann's Encyclopedia of Industrial Chemistry*, 5th edn (eds B. Elvers, S. Hawkins and W. Russey), Wiley-VCH Verlag GmbH, Weinheim, Germany, Vol. A25.
38 Lewis, J.G., Stanley, N.F. and Guist, G.G. (1989) Commercial production and applications of algal hydrocolloids, in *Algae and Human Affairs* (eds A.C. Lembi and R.W. Waaland), Cambridge University Press, Cambridge, UK, Ch 9, pp. 205–236.

39 Baird, J.K. (1994) Gums, in *Kirk-Othmer Encyclopedia of Chemical Technology* Vol. 12, 4th edn (eds J.I. Kroschwitz and M. Howe-Grant), Wiley Interscience, New York, USA.

40 Falshaw, R., Bixler, H.J. and Johndro, K. (2001) *Food Hydrocolloids*, **15** (4–6), 441.

41 Stanley, N.F. (1990) Carrageenans, in *Food Gels* (ed. P. Harris), Elsevier Applied Science, London, UK, pp. 79–120.

42 White, C.A. and Kennedy, J.F. (1983) *Bioactive Carbohydrates in Chemistry Biochemistry and Biology*, Ellis Horwood, Chichester, UK.

43 Lequestel, J.Y., et al. (1995) *Int. J. Biol. Macromol.*, **17** (3–4), 161.

44 Thomas, W.R. (1997) Carrageenan, in *Thickening and Gelling Agents for Food*, 2nd edn (ed. A. Imeson), Springer, New York, USA, pp. 45–59.

45 Imeson, A.P. (2000) Carrageenan, in *Handbook of Hydrocolloids* (eds G.O. Phillips and P.A. Williams), Woodhead Publishing Limited, Cambridge, UK, pp. 45–59.

46 Michel, G. (2006) *Appl. Microbiol. Biotechnol.*, **71** (1), 23.

47 Baird, J.K. (1994) Gums, in *Kirk-Othmer Encyclopedia of Chemical Technology*, Vol. 12, 4th edn (eds J.I. Kroschwitz and M. Howe-Grant), Wiley Interscience, New York, USA.

48 Villanueva, R.D., Hilliou, L. and Sousa-Pinto, I. (2009) *Bioresour. Technol.*, **100** (9), 2633.

49 Uy, S.F., et al. (2005) *Car. Res.*, **340** (7), 1357.

50 Felde, I.F.v.d. and De Ruiter, G.A. (2002) Carrageenan, in *Biopolymers Polysaccharides II: Polysaccharides from Eukaryotes* (eds E.J. Vandamme, S.D. Baets and A. Steinbüchel), Wiley-VCH Verlag GmbH, Weinheim, Germany, pp. 245–274.

51 McHugh, D.J. (2003) Carrageenan, Chapter 7 in FAO Fisheries Technical Papers no. 441: *A Guide to Seaweed Industry*, FAO, Rome, Italy.

52 McHugh, D.J. (2002) Prospects for seaweed production in developing countries, in Fisheries Circular C968, FAO, Rome, Italy.

53 Remminghorst, U. and Rehm, B.H. (2006) *Biotechnol. Lett.*, **28** (21), 1701.

54 Onsøyen, E. (1997) Alginates, in *Thickening and Gelling Agents for Food* 2nd edn (ed. A. Imeson), Springer, New York, USA, pp. 22–44.

55 Brownlee, I.A., et al. (2005) *Crit. Rev. Food Sci. Nutr.*, **45** (6), 497.

56 Moe, S.T. et al. (1995) Alginates, in *Food Polysaccharides and Their Applications* (ed. A.M. Stephen), CRC Press, New York, USA, pp. 245–286.

57 Ertesvag, H. and Valla, S. (1998) *Polym. Degrad. Stab.*, **59** (1–3), 85.

58 Gutowska, A., Jeong, B. and Jasionowski, M. (2001) *Anat. Rec.*, **263** (4), 342.

59 Zimmermann, H., Shirley, S.G. and Zimmermann, U. (2007) *Curr. Diab. Rep.*, **7** (4), 314–320.

60 Thomas, S. (2000) *J. Wound Care*, **9** (2), 56.

61 Qin, Y. (2005) *Int. Wound J.*, **2** (2), 172.

62 Mandel, K.G., et al. (2000) *Ali. Pharm. Ther.*, **14** (6), 669.

63 Wong, T.Y., Preston, L.A. and Schiller, N.L. (2000) *Annu. Rev. Microb.*, **54**, 289.

64 Hoek, C., Mann, D.G. and Jahns, H.M. (1993) *Algae: An Introduction to Phycology*, Cambridge University Press, Cambridge, UK.

65 Holtkamp, A.D., et al. (2009) *Appl. Microbiol. Biotechnol.*, **82** (1), 1.

66 Cumashi, A., et al. (2007) *Glycobiology*, **17** (5), 541.

67 Daniel, R., et al. (1999) *Car. Res.*, **322**, 291.

68 Ghosh, T., et al. (2009) *Glycobiology*, **19** (1), 2.

69 Pomin, V.H. and Mourao, P.A. (2008) *Glycobiology*, **18** (12), 1016.

70 Hayashi, K., et al. (2008) *Int. Immunopharmacol.*, **8** (1), 109.

71 Kusaikin, M.I., et al. (2004) *Dokl. Biochem. Biophys.*, 396.

72 Berteau, O., et al. (2002) *Glycobiology*, **12** (4), 273.

73 Beress, A., et al. (1993) *J. Nat. Prod.*, **56** (4), 478.

74 Descamps, V., et al. (2006) *Mar. Biotechnol.*, **8** (1), 27.

75 Lee, J.B., et al. (2004) *Chem. Pharm. Buli. (Tokyo)*, **52** (9), 1091.

76 Ponce, N.M.A., et al. (2003) *Car. Res.*, **338** (2), 153.

77 Shanmugam, M. and Mody, K.H. (2000) *Cur. Sci.*, **79** (12), 1672.
78 Zetsche, F. and Vicari, H. (1931) *Helv. Chim. Acta*, 14.
79 Ahlers, F., *et al.* (1999) *Phytochemistry*, **50** (6), 1095.
80 Rozema, J., *et al.* (2001) *J. Photochem. Photobiol.*, **62** (1–2), 108.
81 Hasset, J.M., Jennet, J.C. and Smith, J.E. (1980) Heavy metal accumulation by algae, in *Contaminants and Sediments* (eds R.A. Baker), Ann Arbor Science Publishers Inc., Ann Arbor, MI, USA.
82 Pore, R.S. (1984) *Drug. Chem. Toxicol.*, **7** (1), 57.
83 Kessler, E. (1976) *Plant. Syst. Evol.*, **125** (3), 129.
84 Kessler, E. and Huss, V.A.R. (1992) *J. Phycol.*, **28** (4), 550.
85 Erdman, G. (1960) *Svensk. Bot. Tidskr.*, **54**, 561.
86 Ahlers, F., *et al.* (1999) *Z. Naturforsch.*, **54c** (7–8), 492.
87 Gubatz, S., Luftmann, H. and Wiermann, R. (1998) *Bot. Acta*, **111** (4), 291.

8
Natural Products – Market Development and Potentials

Sybille Buchwald-Werner and Friedrich Bischoff

8.1
Introduction

Natural products have been used around the world since human life began. Cultural groups discovered their own fruits and vegetables, tea mixtures, spices, preservatives, traditional herbal medicines and skin oils obtained from indigenous plants. Traditional knowledge has been passed from generation to generation and is in many cases the basis for industrially-produced natural products seen today. Many traditional uses have been verified by science, leading to the discovery of active compounds and modes of action. For example, rosemary is used as a preservative in the Western hemisphere; ginger is used in India. Today, both have proven antibacterial effects and a high content of antioxidants has been demonstrated. Rosemary extracts are used commercially today particularly as a shelf life enhancer for example, for meat.

This chapter focuses on natural products used in foods and pharmaceuticals. The chapter starts with background information about the natural product market. The second section describes the main functional food concepts and the milestones and hurdles to be met during the natural product development process for foods. This is followed by a summary of established natural food ingredients and their benefits. One established condition-specific functional food ingredient is described in detail and an outlook is given about a potential future for a "superfruit". The third section highlights established and innovative drugs derived from plant-made substances. The difficulties in innovation to exploit the immense knowledge about active ingredients from natural plants is discussed.

8.2
Natural Product Market Overview

The global natural product market, focusing on plant derived material, is divided into five market segments. These segments are: active pharmaceutical ingredients

(APIs) based on a single chemical entity; herbal pharmaceutical products; dietary supplements; functional foods; and cosmetic products. Those segments are not strictly separated. It is interesting to note that the degree of regulation increases towards the chemically defined molecules with clear pharmaceutical activity. Because of this diversity and numerous diverging definitions, market numbers vary dramatically. In a study in 1998 by the Business Communications Company (cited in [1]), the world market of "herbal remedies" was estimated to be at 22 billion US$ and forecast to reach 30 billion US$ in 2002. In a report published in 2009 by the same group, the global market for botanical and plant-derived drugs is expected to increase from 19.5 billion US$ in 2008 to 33 billion US$ in 2013 [2]. The range is in line with a study of the European commission estimating the worldwide market for "herbal supplements" at 20 billion US$, and other analysts put the tag at 15 billion US$ for the global market for "herbal supplements" [3]. Gruenwald [4] sees the global herbal pharmaceutical market worth approximately US$ 44 billion to be the biggest market segment for natural products, followed by the global herbal cosmetic market at US$ 14 billion and the global herbal dietary supplement market at US$ 11 billion. The herbal functional food market today has sales of US$ 4 billion sales, but is expected to demonstrate the highest growth rates for natural products in the future [4].

8.2.1
Market for Extracts

The natural products used today can be divided into two broad groups. The first group contains traditionally used natural products, mainly sold as herbal medicines or dietary supplements, like ginkgo, valerian, garlic or echinacea to name just a few examples. These natural products have been well established in the market for many years. Around the year 2000, two factors led to a significant reduction in the volume of sales. First, the public health system in some European countries was changed and some herbal products were excluded from reimbursement by public health insurances. Secondly, negative news about non-proven product claims and adulterations stopped consumers from buying products like St. John's wort and kava-kava, particularly in the US and German market. For instance, the German market which is the biggest single European market for herbal remedies, declined from 2.8 billion in 2003 to 1.9 billion US$ in 2005 [4]. Today these extracts have recovered to a stable annual sales figure between US$ 5–15 million in the US market (Figure 8.1). The second group contains exotic natural products, "superfruits" which are mainly sold in beverages or dietary supplements. Top selling products today are noni and mangosteen (Figure 8.2).

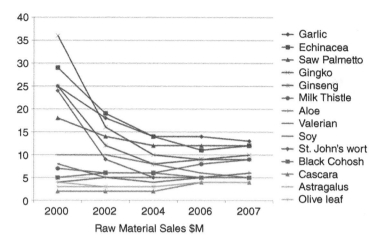

Figure 8.1 Traditional natural products sales in million $ in US from 2000–2007.

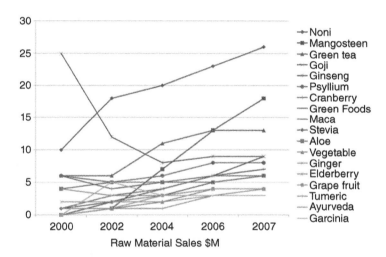

Figure 8.2 Fashionable natural products sales in million $ in US from 2000–2007.

8.3
Natural Products in Food and Functional Food

8.3.1
Functional Food Concept Based on Natural Products

Wellness ingredients (e.g., aloe vera), ingredients with nutritional benefits (e.g., vitamins) and condition-specific ingredients (e.g., plant sterols) with scientifically proven effects are used in functional foods. Products with different development complexity can be differentiated. Consumer products may include well-established, safe ingredients in new food matrices, or established ingredients differentiated by exotic taste or innovative package design. These products are mainly positioned as so-called wellness products. Possible benefits are, for example, calming, energizing, beauty from within or detoxification. Other consumer products contain ingredients with scientifically proven efficacy supported by a special taste or health benefit. These products are called condition-specific functional foods. Benefits are linked to heart heath for example, cholesterol reduction, gut health for example, probiotics, weight management, for example, energy expenditure. Increased innovation impact is coupled with higher development costs and a longer development time (Figure 8.3).

Condition-specific natural products which serve consumer demands for health prevention and reduce health care costs are notably gaining market share. These products target consumer groups with a particular health concern or a specific age profile. The main health concerns are weight management, cardiovascular health, joint health, bone health, brain health. All main health concerns can be aligned to the aging process. The level of good health experienced by aging individuals depends on the interaction between their genetic makeup, the environment and lifestyle choices . Diet and nutritional intake are among the most modifiable lifestyle factors. Several natural products such as antioxidants or other plant-based substances have a proven effect in support of the desired preventative effects, to preserve function and to enhance health throughout life. Many more ingredients are currently under development. Table 8.1 gives an overview of the main established functional food concepts and natural products.

8.3.2
Natural Product Development

Consumers believe in the healthy effect of foods or beverages when they contain natural ingredients which are traditionally known and generally recognized as safe and healthy. Everything which is "new" and does not have a traditional use needs to have the support of extensive marketing and PR campaigns to achieve consumer awareness and understanding. Different trends can be observed for the US and European markets. In the USA "superfruits", which are exotic ingredients combined with a taste experience and an easy to understand benefit like "high in antioxidants", show market success. These products are mainly formulated as drinks and supported are by fair trade and bioethical labeling. In Europe, particularly in the UK, herbal products referring to traditional knowhow from other cultures like traditional Chinese medicine (TCM) or ayurveda are fashionable.

In the past, commercially available natural products relied on traditional knowledge about the use and the preparation of the plant material. Raw materials were collected from the wild by knowledgeable people. Only a little was known about the chemical composition of the plants and their consistency. Identification macroscopically and microscopically was sufficient to identify the species and exclude adulterations. Over time, the growing demand for selected raw materials could no

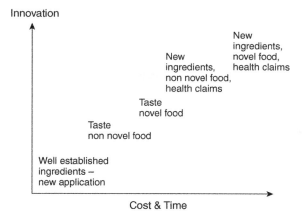

Figure 8.3 Innovation impact in functional foods referring to cost and time of the development.

Table 8.1 Functional foods benefit areas and frequently used ingredients.

Health areas	Benefits	Established ingredients
Antioxidant	Antioxidants	Superfruits, berry extract, carotenoids
Beauty	Antioxidants, anti-aging, moisturizing	Plant extracts (e.g., aloe vera, cocoa, green tea, grape seed) fatty acids (e.g., evening primrose oil, fish oil)
Bone health	Anti-osteoporosis	Minerals, vitamin K, vitamin D, plant extracts (phytoestrogens, genistein)
Brain	Calming, sleep promotion, improved cognitive function	Plant extracts (e.g., passion flower, lemon balm)
Eye	Improved vision	Carotenoids (e.g., lutein), plant extracts (e.g., bilberry)
Gut	Probiotics, fiber	Bacteria, plant extracts (e.g., aloe vera), inulin
Heart	Cholesterol reduction, blood pressure balancing, circulation, healthy vessels	Phytosterols, peptides, fibers, plant extracts (e.g., cocoa bean, olive leaf, grape seed, tomato)
Joined health	Anti-inflammation, anti-pain	Glucosamine, plant extracts (e.g., lemon verbena)
Sport	Energy	Proteins, amino acids, plant extracts with caffeine for example, guarana
Weight management	Reduction in body fat, satiety, low gastro-intestinal activity, thermogeneses, increased fat burning, reduced fat absorption	Conjugated linoleic acid, plant extracts (e.g., garcinia, green tea, green coffee bean, opuntia), pinolenic acid, potato starch, palm & oat oil, orlistat

longer be served by wild collection and cultivation had to be established. It became necessary to understand the quality, composition and subspecies of material collected from the wild to obtain comparable raw material qualities by cultivation. Subsequently, specifications for natural products advanced and did include information about the composition of the product. In addition, monitoring for contaminants like other species, pesticides and environmental toxins was established. Today plant extracts and their preparation are standardized to ensure consistent composition, safety and efficacy of the natural product.

Globally, new product developments for condition-specific functional foods are targeted. These projects are complex and long term. Overall development project timelines range from 3–5 years, mainly depending on nonpredictable timeframes for regulatory approvals. GRAS (generally recognized as safe) and a Food and Drug Administration (FDA) approved health claim is necessary in USA. In Europe, all natural products which have not been used in foods before 1997 have to get a novel food notification, which notifies that the product is safe. Health claims are evaluated by the European Food Safety Agency (EFSA). The health claim approval process is new and the first claims were only approved in 2009. More than 1000 applications are waiting for approval. Therefore, the timing for evaluation of a new application is more or less unpredictable. Regulatory approvals to demonstrate product quality, safety and scientifically substantiated health are getting more and more complex. Consequently this leads to limited market access for small companies which do not have the capacity and funding to go through the approval process. As a result fewer innovations are being seen on the market. Table 8.2 gives an overview about the development steps and timeframe for a functional food natural ingredient.

In addition, product development is challenged by the availability of sustainable raw materials. Sustainable raw material availability and acceptable raw material prices are both at risk because of several factors, such as increasing weather impact, lack of geographical diversification and increasing transport costs. Enhanced environmental policies and energy efficiency programs resulted in closure of small manufacturers who are not able to meet these new requirements, particularly in China. Sustainable raw material availability is today the key requirement for successful new product development. Modern consumers are sensitive about environmental protection and social responsibility. First choices are domestic or regional products to reduce the carbon footprint. For other natural products fair trade and bioethical standards are gaining importance. Benefit sharing for the use of genetic resources and traditional knowledge related to indigenous plants will become a unique selling point for future natural products.

8.3.3
Food Applications of Natural Products

Sustainable raw material sources, cost-efficient production process, demonstrated safety and efficacy and regulatory approvals are basic requirements but no guarantee of market success. Several success factors can be identified. The first

Table 8.2 Timelines for natural product development for functional foods.

	Development milestones	Year 1	Year 2	Year 3	Year 4	Year 5
1	**Feasibility study** – Raw material availability – Traditional knowhow – Theoretically safety evaluation – Regulatory status evaluation for target markets – IP evaluation, freedom to operate – Potential market evaluation based on targeted product benefits and application	▬				
2	**Phytochemical identification** – Plant composition – Key compounds, actions – Analytical method development	▬				
3	**Sustainable raw material sources** – Evaluation of raw material from different regions and suppliers – Identification of the representative plant composition and quantification – Evaluation of capacity and costs – Set up supply chain	▬				
4	**Process and analytical development** – Development of extraction process at lab scale – Development of analytical methods		▬			
5	**Proof of concept** – *In vitro*, animal study		▬			

Table 8.2 Continued

6	Safety vitro	
7	IP evaluation, first patents	
8	Extraction at pilot plant	
9	Safety in animals	
10	Food application trials	
11	Market – regulatory strategy – Targeted markets – Necessary regulatory approvals – Target health claims – Estimated business plan	
12	Safety efficacy *human studies*	
13	Regulatory approvals	
14	Production set up	
15	Marketing strategy, launch plan	

one is the taste. Functional foods are foods and not medicinal products. An unpleasant taste or flavor of an ingredient is not tolerated by consumers. Most of the secondary plant ingredients are bitter or spicy and require taste masking and encapsulation techniques, which enhance the taste by maintaining the bioavailability.

The second success factor is the fit of the technological properties of the natural ingredient to the right food matrix. For example, natural products with beneficial effects for sportive people, like isotonic blends or ingredients to support joint health are preferably formulated as beverages. Suitable ingredients are water soluble or technologically modified to be water dispersible as well as heat stable to allow for pasteurization treatment. The link between the natural product and suitable food matrices has to be established during the first phase of product development. Key technological parameters are solubility, dispersibility and stability. Vegetable oils are often used in the ester form to improve their ability to be incorporated into food matrices. Esters have a significantly reduced melting point compared with the oil. Therefore esters are more suitable for incorporation into semi-solid food matrices like spreads and yoghurts. Grinding is another option to make insoluble natural products suitable for incorporation into semi-solid or liquid food matrices, for example, cryogenic grinding is used to obtain particle sizes of 1–2 micrometer.

The third success factor is the selection of the right food formulation. A suitable food matrix should have a proven track record to transfer the specific product benefits to the consumer. Nevertheless an effective communication strategy is essential. Innovative products require a communication strategy which translates the science behind the products simply and effectively, thus enabling consumers to come to their own decisions and select exactly what they want. Consumer education and credibility obtained from adequate information are key topics to be included in strategies for the marketing of future functional food products with proven efficacy. Table 8.3 gives an overview about functional foods benefits, mainly

Table 8.3 Functional foods benefit areas and frequently used food matrices.

Benefit area	Frequently used food matrices
Brain health	– Beverages
	– Cereals, bars
Beauty inside	– Beverages
	– Smoothies, dairy drinks
Gut health	– Dairy products
Heart health	– Dairy products
	– Spread
Joined / Sports	– Beverages
	– Bars
Weight	– Dairy products

used food matrices and technological requirements for the ingredient. The last factor is that consumers are generally looking for convenience. Therefore beverages, on-the-go-shots as well as bars may have a higher chance of market success than other functional foods products.

8.3.4
Success Story Plant Sterols for Heart Health

The World Health Organization (WHO) estimates that worldwide approximately 17 million people die of cardiovascular disease (CVD) every year. In Europe 50% of all deaths are related to CVD. A wide range of potentially reversible risk factors for CHD has been identified. These include: smoking, nutritional habits leading to increased serum lipids, and physical inactivity [5, 6]. Considerable attention has been paid to the association between blood cholesterol levels and the development of CVD. Cholesterol is a lipid substance that is formed by the body itself and absorbed from food. It is the initial building block for various hormones and membranes. If its serum concentration is above normal levels for an extended period, cholesterol can form deposits on the inner walls of the vessels, which may lead to arterioscleroses and occlusions resulting in CVD, heart attack or stroke. Cholesterol plasma levels can be lowered by reducing saturated fatty acids and increasing polysaturated fatty acids (e.g., fish oil) and soluble fiber in the daily diet. However, experience has shown that it is difficult to change lifestyle and eating habits. Functional foods offer consumers the opportunity to balance their nutrition with food elements that provide nutritional value plus additional benefits to maintain health.

Plant sterols are well established functional food ingredients. They are added to a range of foods such as spreads, yoghurts and drinks which are consumed for their ability to lower cholesterol levels, particularly low density lipoprotein (LDL) and by inference the risk of CHD. The average scale of reduction of LDL, around 5–15%, is well above that which can be achieved solely by manipulating dietary fat composition [7, 8]. Phytosterols (plant sterols) are naturally occurring compounds with similar chemical structure and biological functions as cholesterol [9]. They are mainly found in plant oils, seeds, nuts, vegetables and fruits. The most abundant plant sterols are sitosterol, campesterol and stigmasterol [10].

The daily dietary intake of plant sterols in food is around 160–400 mg, but much earlier in human evolution, intake would have been much higher, up to 1 g per day. A strictly vegetarian diet may provide up to 600–800 mg per day. Plant sterols have a clear cholesterol-lowering effect, either in their free or esterified form. [11] Clinical studies showed that the intake of 1–3 g phytosterols per day can significantly decrease the LDL levels by 5–15%. The first effects have been demonstrated after 1–2 weeks. Phytosterols are specially recommended for a person with slight to moderately raised LDL levels and who wishes to actively reduce this level via his or her diet. Health organizations like the American Heart Association recommend phytosterols as one option to reduce cholesterol levels and the risk of CVD.

Commercially available phytosterols and phytostanols are isolated from tall oil or vegetable oils such as soybean oil, rape seed oil, sunflower oil or corn oil. Although commercial preparations comprise varying proportions of individual sterols, they are generally referred to as phytosterols or phytostanols depending on which type is predominant. Compelling scientific evidence demonstrating the safety and benefits of plant sterols and plant sterol esters has led to regulatory approvals. Food products containing plant sterols and plant sterol esters are classified as novel foods in Europe and have been approved by the European Commission [12]. They are generally recognized as safe in USA (FDA–GRAS approval) and also approved in Asia Pacific as novel foods (e.g., in Japan, Korea, Australia and China official approval has been given).

8.3.5
Baobab–The New Superfruit

Baobab is a new exotic food and beverage ingredient with a unique tangy taste which is ethically and sustainably produced. Baobab trees (*Adansonia digitata* L.) are distributed widely across Southern Africa where the main producing countries include Malawi, Zimbabwe, Mozambique and Namibia. The baobab tree (Figure 8.4) bears large ovoid fruits which have a hard woody shell. A number of seeds are contained within the shell and embedded in a powdery pulp (Figure 8.5). The pulp is separated from both the seed and shell to produce a fine, white to pinkish-brown powder with no odor, but an acidulous taste. In addition to the pulp, the baobab seeds can be used to produce oil.

The baobab fruit is used traditionally as a nutritional ingredient as well as a medicine [13]. A cool, refreshing drink is made using the pulp which contains significant quantities of macronutrients and micronutrients [14, 15]. The fruit pulp is also known for its high mineral and vitamin content and contains high level of carbohydrates, vitamin C, calcium, potassium, thiamine and nicotinic acid [16]. Due to the fiber content it is a potential pre-biotic ingredient. The baobab fruit powder is used as a raw material for baobab products to be used in foods and functional foods. Traditionally used in drinks is the natural fine white pulp powder, which is mechanically separated from the baobab fruit. In addition it is used as a performance ingredient to modify viscosity and or moisture of food matrices (pectin). Possible food applications include dairy products, smoothies, and cereal and fruit bars. Through receiving EU Novel Foods and FDA–GRAS approval, baobab fruit powder has proved to be safe as a food ingredient. In order to serve the beverage market it was necessary to modify the natural pulp powder to obtain a clear water-soluble, nonviscous natural product. Through this process, insoluble components are separated as well as the pectin. The pectin is enzymatically removed and the powder is spray dried to produce the depectinized baobab fruit powder. The depectinized extract obtained is clear, water-soluble, and a perfect ingredient for beverages. Depectinized baobab extract is also approved as a Novel Food in the EU [17].

Figure 8.4 Baobab tree.

Figure 8.5 Baobab fruit open to reveal the chunks of powder.

Baobab fulfils nearly all the requirements for successful market introduction into the functional food sector. It is a natural product with a pleasant, refreshing taste and is rich in antioxidants and minerals. It can be formulated in beverages and bars which serve as high convenience products for consumers. The product is safe and has main regulatory approvals. In addition to its good taste and nutritional profile, the supply chain is sustainable. Global best-practice forestry techniques are employed to ensure that the abundant supply of baobab in Southern Africa is managed sustainably in the long term. Furthermore, baobab is fair- and bioethically-traded and producers share part of their income with their villages to fund community projects. The only hurdle is that baobab has come across is its exotic origin and that it is not well known to Western consumers. Extensive marketing and PR campaigns are necessary to establish consumer awareness and knowledge of this product. Initial campaigns have been launched and Mintel, a highly acclaimed Market Research Company, said that baobab has started its career as next superfruit in 2009.

8.4
Natural Products for Pharmaceutical Applications

8.4.1
Existing Products and Innovation

Traditionally, there has been a clear distinction between botanical extracts based on traditional knowledge, and chemically defined molecules extracted and purified (and often chemically modified). As it has already been discussed, the worldwide market value for herbal medicinal products ("botanicals") is around 20 million US$. In the EU, Germany (27%) is the largest market before France (24%) [4]. However, changes in regulations have not only led to a decrease of the market, but raised the bar dramatically for new products to be introduced to the market. In principle, the market of herbal medicinal products is restricted to those products for which decades of traditional use have shown no adverse effects. Monographs fix not only the nature of solvent (mainly water and ethanol), but also the amount of solvent by a drug-to-extract-ratio (DER). These regulations even prevent extracts from being improved by increasing the selectivity of the process (provided that the nature of the active principles is known).

The situation in the USA is different. In June 2006, the FDA published the "Guidance for Industry–Botanical Drug Products", which provides a path for submitting investigational new drug applications (NDA) for botanical drug products. Consequently, it is not surprising that the first botanical drug product having undergone clinical trials was approved already 2006 by the FDA for the US market, but approval for this innovative drug took three and four years longer in Germany and Austria, respectively. Polyphenon®E Ointment, which was developed by Medigene AG (Germany) and Epitome Pharmaceuticals (USA), was the first botanical drug approved under the 1962 Drug Amendment which requires that the safety and effectiveness must be proved prior to the marketing of a drug. The drug product, named Veregen®, contains catechins that are extracted by water from green tea leaves and subsequently purified by chromatography. Veregen® is specifically indicated for the topical treatment of external genital and perianal warts caused by human papilloma virus.

Regulatory hurdles for drugs containing only one chemical entity are also high, but have been established for decades. The list of important drugs comprises numerous drugs based on natural products, for example antibiotics. However, only four blockbusters (drugs with sales greater than one billion US$/a) are derived from plant-made molecules: Oxycontin® from thebaine, Spiriva® from scopolamine, Taxotere® from deacetyl-baccatin and Tamiflu® from shikimic acid. Most established products derived from plants have lower annual sales. Table 8.4 lists established drugs derived from alkaloids extracted from natural sources. The comprehensive list nicely demonstrates the diverse therapeutic applications of this structural class of natural molecules. According to a recent report, respiratory problems represent the largest application of plant-derived drugs in 2005, accounting for 26% of total sales of plant-derived drugs [18]. Interestingly, cancer treatment is expected to become the largest application of plant-derived drugs by 2011, with 24% of the market.

However, there are very few show cases for new leads from plants. This might be due to the trend for targeted screenings within the pharmaceutical industry. Once a new target is identified and validated by academia, a race starts in the laboratories of pharmaceutical companies to set up molecular screening models for this particular target. When the target screen is established, huge libraries comprising hundred-thousands of compounds are screened against those targets. Only a small fraction of those substances are natural compounds extracted from plants, whereas most of the chemical compounds have been created by combinatorial synthesis. In contrast to natural compounds, the chemically synthesized compounds have already been selected for features like good solubility and favorable pharmacokinetic profiles. It is not surprising that after a first round of screening the natural compounds are dropped because of their poor characteristics. In addition, they do not offer an easy access to standard chemistry to improve their

Table 8.4 Active pharmaceutical ingredients derived from natural alkaloids.

Name of API	Source for extraction	Therapeutic use
Atropine	Different solanaceae, but also by synthesis	Anticholinergic (eye dilation)
Hyoscyamine and scopolamine	Different solanaceae	Anticholineric
Scopolamine-N-butyl-bromide, cimetropium / butropium bromide	Semi-synthesis using scopolamine from different solanaceae species	Anticholinergic (gastro-intestinal)
Tiotropium bromide	Semi-synthetic	Anticholinergic (chronic obstructive bronchitis, asthma)

Table 8.4 Continued

Name of API	Source for extraction	Therapeutic use
Tropisetrone	Semi-synthetic	Anti-emetic
Cocaine	*Erythroxylum coca*	anesthetic
Morphine	*Papaver somniferum*	Pain treatment
Codeine	Semi-synthetic from extracted morphine	Analgesic
Oxycodone ...	Semi-synthetic from extracted thebaine	Analgesic
Irinotecan, topotecan, belotecan	Semi-synthesis using camptothecin extracted from *Camptotheca accuminata* and *Nothapodytes foetida*	Cancer treatment
Physostigmine	*Physostigma venenosum*	Cholinergic
Reserpine	*Rauwolfia serpentina*	Anti-hypertensive
Yohimbine	*Pausinistalia yohimbe*	Aphrodisiac
Vinblastine, vincristine, vindesine, vinorelbine, vinfluinine	*Catharantus roseus*	cancer treatment
Quinine	*Cinchona* species	Malaria treatment
Colchicine, thiocolchicine	*Colchicum autumnale*	Gout treatment
Emetine	*Cephaelis ipecacuanha*	Anti-amoebic, anticough
Galanthamine	*Narcissus pseudonarcissus*, but also by synthesis	Cholinesterase inhibitor (Alzheimer)
Tubocurarine	*Strychnos toxifera*	Skeletal muscle relaxant
Nicotine	*Nicotiana tabaccum*	smoking cessation therapy
Capsaicin	*Capsicum* species	Analgesic
Pilocarpine	*Pilocarpus jaborandi*	Cholinergic (eye pressure reduction)
Ergotamine, dihydroergotamine (DHE)	Extracted from *Claviceps purpurea* (plant parasitic fungus), DHE is semi-synthetic	Migraine treatment
Cabergolin	Semi-synthetic	Lactation suppression, Parkinson disease treatment
Nicergoline	Semi-synthetic from lysergol (Ipomoea) or from fermentation	Dementia treatment
Trabectedin	Semi-synthetic from *Ecteinascidia turbinata* (sea squirt) or micro-organism	Cancer treatment

properties. The development timelines are thought to be very long which is exemplified by the drug examples mentioned below. In cases where generic competition starts even before the expiry of the patent, long development cycles are not acceptable for pharmaceutical companies. Therefore, natural compounds are often found by pure incidence or a non-targeted approach and are further developed in academia. A recent example is omacetaxine mepesuccinate which is described in more detail below.

Another type of new natural products stems from an incremental development. A good example is the development of several vinca alkaloids. Vinflunine is a new derivative developed by Pierre Fabre (France) effective in final clinical trials [19]. It is a modification of vinorelbine which was launched in 1989. Vinorelbine was an improved derivative due to its selectivity for mitotic microtubules (which are essential for rapidly dividing cancer cells) versus axonal microtubules (which are important for the nervous system). Neurotoxicity was one of the limiting factors of the older vinca alkaloids (vinblastine, vincristine and vindesine). Vinorelbine is mainly used in lung cancer. Vinflunine has been recommended for approval by the Committee for Medicinal Products for Human Use (CHMP) for transitional cell carcinoma of the urothelium. Despite the incremental innovation, combined sales of all different vinca alkaloids have gradually declined during the last twenty years to approximately 100 million € per year. A substantial amount of extraction and semi-synthesis is carried out in France regardless of increasing generic competition.

8.4.2
Terpenes: Paclitaxel (Taxol) – The Posterchild

Taxol was isolated as an anticancer substance in 1971 in the laboratory of Wall and Wani from the bark of the pacific yew tree (*Taxus brevifolia*). However, the content is very low ($0.1 \, g \, kg^{-1}$). To fulfill its supply needs, the US pharmaceutical company Bristol-Myers-Squibb collaborated with the US company Hauser, based in Boulder (Colorado), to extract large amounts of yew tree bark. Due to its slow growth, the tree had nearly become extinct. The supply situation dramatically improved since 1994 when paclitaxel was produced using the pre-cursor 10-deacetylbaccatin (10-DAB). This compound is isolated from needles of different yew varieties in a sustainable manner. Bristol-Myers-Squibb arranged to obtain 10-DAB from Indena, based in Milan (Italy).

Paclitaxel first received FDA approval in 1993 for the treatment of ovarian cancer. Approvals for treatment of breast cancer and lung cancer followed. The branded drug (Taxol®) generated 1.6 billion US$ in sales in 2000 before paclitaxel became available as generic. The Canadian Forest service estimated the demand for paclitaxel at 800–1000 kg worldwide.

Docetaxel is another semi-synthetic derivative developed initially by Rhone-Poulenc (now Sanofi-Aventis, France). The replacement of the acetate ester at carbon 10 of paclitaxel for a hydroxyl group makes docetaxel more water-soluble. The branded drug (Taxotere®) was initially approved in 1996 for breast cancer.

It is currently approved also for ovarian cancer, non-small cell lung cancer (NSCLC), advanced gastric and prostate cancer. Taxotere reached 2177 Million € sales in 2009, but the market is predicted to shrink as patents expired in 2010.

Due to the "opening" of the market, a number of different companies have entered the market of extracting taxanes. In addition, efforts to increase the taxane content of yew were successful. Large plantations have been established for example, in Canada and USA. Together with improved purification processes, those efforts have helped to decrease production costs from approximately 300 000 US$ per kg paclitaxel to prices below 100 000 US$ per kg.

8.4.3
Alkaloids: A Group of Potent APIs

8.4.3.1 Camptothecin-derived APIs

Camptothecin is another discovery from laboratory of Wall and Wani in the early 1960s. The substance was initially isolated from the Chinese tree, *Camptotheca acuminata*, but was found later in higher amounts (up to 0.4%) in another Asian plant, *Nothapodytes foetida* [20]. However, both plants are still used for extraction mainly carried out in China and India. The annual quantity may reach 1000 kg camptothecin extracted and purified.

It took almost 40 years of intensive research to develop effective semi-synthetic compounds with less side effects. Irinotecan (7-ethyl-10-[4-(1-piperidino)-1-piperidino]-carbonyloxy-camptothecin) was developed by Yakult Honsha and first approved in Japan in 1994. It is approved as first-line treatment of metastatic colorectal cancer in combination with 5-FU/leucovirin, but is also used to treat lung, cervical and ovarian cancer. The world-market for irinotecan is around 400 kg per year according to IMS Health data. Sales of irinotecan by Yakult Honsha, Aventis and Pfizer peaked in 2005 at approximately three billion US$. After irinotecan lost patent protection, the prices dropped significantly and Fresenius Kabi (Germany) acquired the largest part of the generic market, according to IMS Health data.

As irinotecan, topotecan (9-dimethylaminomethyl-10-hydroxy-camptothecin) is more water-soluble and less toxic than camptothecin. Topotecan was developed by GlaxoSmithKline (GSK) [21] and launched in 1996 for the second-line treatment of metastatic ovarian cancer. The FDA approved topotecan later for relapsed small-cell lung cancer and recurrent cervical cancer, too. But sales reached only 72 million US$ in 2009, according to GSK earnings statement. Belotecan, developed by Chong Kun Dang, was only launched in Korea in 2004 for ovarian and small cell lung cancer. A liposomal formulation for the US market is still in phase II clinical trials.

Many other derivatives have been under development by several companies, but until today, none of them has been effective enough to be approved. Daiichi Sankyo (Japan) decided to discontinue development of exatecan after a large clinical phase III study in 2004. Lipophilic derivatives of camptothecin are still being studied: cositecan by the company Bionumerik (USA) is in phase III clinical phase;

gimatecan by the company Sigma-Tau (Italy) is in phase II. For the years to come it is not expected that the demand of camptothecin as starting material for the synthesis of anticancer drugs will significantly rise.

8.4.3.2 Omacetaxine (Homoharringtonine)

Omacetaxine mepesuccinate has been developed from the natural cytotoxic compound homoharringtonine. The antitumor properties of alkaloids extracted from *Cephalotaxus* were first discovered by Chinese investigators. In the 1970s, Powell et al. [22] isolated a series of esters of cephalotaxine with anticancer activity and elucidated the structures. For homoharringtonine, it was shown that protein synthesis is inhibited through its suppression of substrate binding to the receptor site on the 60S ribosome subunit [23]. In contrast to extracting bark from *Cephalotaxus harringtonia*, a more efficient way to produce the API (omacetaxine mepesuccinate) was found ten years ago: The alkaloid cephalotaxine is extracted from dry leaves of *Cephalotaxus* trees and subsequently the ester group is added to yield semi-synthetic homoharringtonine. The clinical development of homoharringtonine in chronic and acute myelogenous leukemia (CML and AML) paused when tyrosine kinase inhibitors like imatinib mesylate (Gleevec) became available in late 1990s. But due to its independent mechanism, interest in homoharringtonine resurrected due to the occurrence of cancer cells resistant to tyrosine kinase inhibitors in 20–30% of patients. After conducting clinical trials [24], the company Chem-Genex (California, USA) has filed omacetaxine for treatment of CML and is awaiting final approval by the FDA in 2010.

8.4.3.3 Opiate Alkaloids

Opium and poppy straw are the raw materials from which alkaloids, such as morphine, thebaine, noscapine and codeine, are extracted for medicinal purpose. Opium raw material is the latex obtained by making incisions on the green capsules of the poppy plants. The latex turns into a dark resin called raw opium. Instead of opium, "concentrate of poppy straw" (CPS) is mainly imported which is a dried residue from the extract of poppy straw. Morphine is the main alkaloid in Indian opium and Turkish CPS. Since the mid-1990s, France and Australia started the commercial cultivation of opium poppy with high thebaine content in response to the increasing need as a starting material for the synthesis of oxycodone. The large increase in oxycodone use in the USA resulted from the approval in 1995 of a high-dose, extended-release drug formulation. The group of Purdue Pharma, with affiliate Mundipharma, reported for its finished dosage form Oxycontin® 2.5 billion US$ annual sales in 2007 and 2008, respectively. Approximately 75 tons of oxycodone are manufactured worldwide representing approx. 4 billion US$ annual sales in finished dosage forms. 51.6 tons oxycodone have been consumed in the USA in 2007 according to the International Narcotics Control Board. However, behind the 2.5 billion US$ oxycodone business is only a value of approximately 5.85 million US$ for the crude extract, that is, CPS. Interestingly, the producers seemed to have coped well with the sky-rocketing consumption as the import price for thebaine-rich

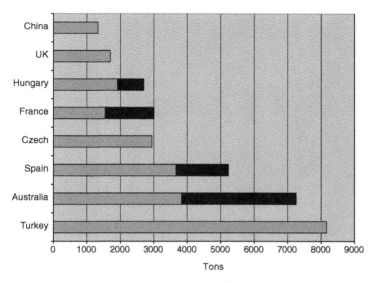

Figure 8.6 Quantity (in tons) of poppy straw rich in morphine (shaded) or thebaine (black) produced in 2007 in the listed countries (Source of data: NCBI statistics).

CPS decreased from $825 per kg in 2001 to $500 per kg in 2007, according to a recent report from the US Drug Enforcement Administration. Production quantities for morphine and thebaine equivalents are shown in Figure 8.6. The amount of thebaine from India is due to the fact that thebaine is also present in Indian opium at approximately one sixth the level of morphium. According to NCBI statistics for the year 2007, 25 294 tons of poppy straw rich in morphine and 6464 tons of poppy straw rich in thebaine were produce worldwide for extraction of alkaloids.

Not surprisingly, the extraction, purification and production of semi-synthetics is strictly regulated and monitored. For instance, all narcotic substances are controlled under schedule II or schedule III in the USA. Therefore, only a limited number of companies is involved in extraction and semi-synthesis. In Figure 8.6 and Figure 8.7, an overview is given about countries involved in production of raw materials and extraction, respectively. It is obvious that most of the poppy straw is extracted in the countries growing poppy (Australia, France and Spain). However, Australia exports most of its concentrated extract (CPS) to the USA for manufacturing thebaine and finally oxycodone. Only 10 168 kg thebaine were produced in 2007 in Australia and no oxycodone. For Spain, the amount of poppy harvested, the amount of thebaine extracted (ATA) and the amount of thebaine manufactured fit together. Apparently, thebaine from Spain is exported. France seems to convert most extract into thebaine and finally into oxycodone. India and Japan produced thebaine exclusively by extracting opium. In 2007, 4140 kg thebaine have been extracted as a by-product from opium in the USA, whereas most of the thebaine manufactured in the USA and the UK is purified from concentrated extracts

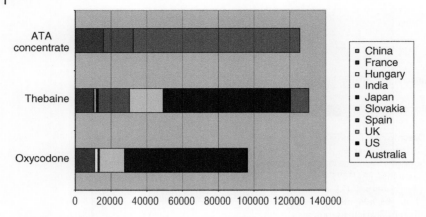

Figure 8.7 The amounts of anhydrous thebaine alkaloid (ATA) concentrated poppy straw extract, thebaine and oxycodone produced in each country (in kilograms).

mainly imported from Australia. Besides oxycodone production in the USA, UK and France, small amounts of oxycodone are produced in Hungary, Slovakia and Japan (Figure 8.7). This example shows nicely how the different manufacturing steps from raw material production, extraction, purification of the active substance and conversion into the semi-synthetic final drug can be separated between different manufacturers and countries. An even more complicated picture would result from an analysis of codeine as there are two salt forms and different uses: The phosphate and sulfate salt form of codeine are mainly semi-synthetically produced from morphine. Codeine API annual value is still increasing and now close to 200 million US$ (364 tons codeine produced in 2007). As codeine is used in many different combination products, it is difficult to give a value for the final codeine drug product market.

8.5
Summary

It can be concluded that much innovation is going on in the area of natural products. However, it seems that innovation mainly takes place in areas with less regulation or in areas which opened by new regulations for example, the new botanical drug product regulation in the USA. The frontiers between pharmaceuticals, cosmetics and food are becoming fuzzier. Cosmeceuticals and nutraceuticals with health claims are gaining important market shares. Moving away from the classical applications in the pharmaceutical industry, the product cycles are shorter and there is strong pressure on development and production costs. Other chapters of this book describe how these targets can be reconciled by technological improvements.

References

1. Raskin, I., et al. (2002) *Trends Biotechnol.*, **20**, 522.
2. Business Communications Company Inc. (2009) Botanical and plant-derived drugs. Report BCC00102.
3. Glaser, V. (1999) *Nat. Biotechnol.*, **17**, 17.
4. Gruenwald, J. (2008) The global herbs & botanicals market. *Nutraceuticals World*, http://www.nutraceuticalsworld.com/contents/view/13953.
5. Westendorp, R.G.J. (2006) *Am. J. Clin. Nutr.*, **83**, 404. http://www.who.int/dietphysicalactivity/publications/facts/cvd/en/ (accessed 16 November 2009).
6. Mann, J. (2004) Cardiovascular disease, in *Public Health Nutrition* (eds M.J. Gibney, et al.), Blackwell Publishing, Oxford, UK, pp. 317–330.
7. Berger, A., Jones, P.J.H. and Abumweis, S.S. (2004) *Lipids Health Dis.*, **3**, 5.
8. Weidner, C., et al. (2009) *Lipids Health Dis.*, **7**, 35.
9. Piironen, V., et al. (2000) *J. Sci. Food Agric.*, **80**, 939.
10. Moreau, R., Whitaker, B. and Hicks, K. (2002) *Prog. Lipid Res.*, **41**, 457.
11. Nestel, P., et al. (2001) *Eur. J. Clin. Nutr.*, **55**, 1084.
12. European Commission, (2009) Commission Decisions authorising the placing on the market of novel food. http://ec.europa.eu/food/food/biotechnology/novelfood/authorisations_en.htm (accessed 16 November 2009).
13. Chadare, F.J., et al. (2009) *Crit. Rev. Food Sci. Nutr.*, **49**, 254.
14. Osman, M. (2004) *Plant Foods Hum. Nutr.*, **59**, 29.
15. Hassan, A. and Hassan, H. (2008) *Sci. World J.*, **3** (2), 113.
16. PhytoTrade Africa (2008) Baobab Dried Fruit Pulp – An application for Novel Foods Approval in the EU as a food ingredient. http://www.food.gov.uk/multimedia/pdfs/baobabapplicationfinal.pdf (accessed 16 November 2009).
17. FDA (2008) Agency Response Letter GRAS Notice No. GRN 000273. http://www.fda.gov/Food/FoodIngredientsPackaging/GenerallyRecognizedasSafeGRAS/GRASListings/ucm174945.htm (accessed 16 November 2009).
18. Business Communications Company Inc. (2006) Plant-derived drugs: products, technology, applications. Report WA1378188.
19. Bellmunt, J., et al. (2009) *J. Clin. Oncol.*, **27**, 4454.
20. Fuzele, D.P. and Satdive, R.K. (2005) *Fitoterapia*, **76**, 643.
21. Kingsbury, W.D., et al. (1991) *J. Med. Chem.*, **34**, 98.
22. Powell, R.G., Weisleder, D. and Smith, C.R. (1972) *J. Pharm. Sci.*, **61**, 1227.
23. Fresno, M., Jimenez, A. and Vasquez, D. (1977) *Eur. J. Biochem.*, **72**, 323.
24. Quintas-Cardama, A., Kantarijan, H. and Cortes, J. (2009) *Cancer*, **115** (23), 5382–5393.

9
Regulations and Requirements

Guido Ritter and Alwine Kraatz

9.1
Introduction

Natural plant products (or "botanicals") have served as a major source of aromas, food ingredients, food additives and novel foodstuffs for centuries. Only a few food ingredients are chemically synthesized as copies of the natural structure in an industrial scale, vanillin being one of the first examples. But it is not only for foodstuffs, since about half of the pharmaceuticals in use today are derived from natural products. Natural ingredients are defined as those not produced by chemical synthesis, but extracted from biological material, like plants. Plant products, for example, starch, sugar, cacao, tea and coffee, have become so important in our nutrition, that a complete industry with its own extraction/processing technology and regulation is effective today.

The mature food market in developed countries is greedy for food innovations. One of the most important trends by both consumers and industry is the health food sector. Historically, plants have been used as a valuable source of prophylactic agents in the prevention or treatment of chronic diseases; and the scientific understanding of the beneficial health impact of bioactive components of foods has improved significantly in the last decade. Along with the effects of globalization and the increased import of food, botanicals and derived preparations (made from plants, algae, fungi or lichens) with perceived and real health benefits have become widely available on the European market. Examples include ginkgo, garlic, St. John's Wort and ginseng. Such products are typically labeled as natural foods and a variety of claims are made regarding their possible health benefits. They can be bought over the counter in pharmacies, supermarkets, herbalists and via the internet. While most of these products have a long history of use, some concerns exist with regard to quality and safety. These include the risk of chemical or microbiological contamination (particularly in products imported) and the need to ensure that concentrations of bioactive agents are within safe limits. Extraction, processing and marketing of natural plant products ("botanicals") requires the consideration of regulations. This chapter covers the definition, regulations of the application including

Industrial Scale Natural Products Extraction, First Edition. Edited by Hans-Jörg Bart, Stephan Pilz.
© 2011 Wiley-VCH Verlag GmbH & Co. KGaA. Published 2011 by Wiley-VCH Verlag GmbH & Co. KGaA.

labeling and requirements concerning quality and safety issues of natural food products in the European Union (EU).

9.2
Definition of Plant Extracts as Medicines or Foods – A Borderline Issue

There is currently no consistency in the legal status of some botanicals across the EU. In some EU countries, botanical products are marketed as foods, food supplements or incorporated into fortified foods; whereas in other EU countries these preparations are seen as herbal medicines registered by full or simplified registration procedures [1]. Although food and medicinal law clearly allow the use of botanicals both in food and medicinal products, the use of some botanicals in foods is sometimes challenged by national authorities in their aim to defend the end-consumer from misuse of wrong self-medication.

9.2.1
Medicines

The member states of the EU interpret the definition of a medicine in the context of products containing botanicals differently, due to varied interpretations of the terms "presentation" and "function" given in the EU legislation (Directive 2001/83/EC as amended by Directive 2004/27/EC) [2], the definition of a medicinal product:

> "Any substance or combination of substance presented as having properties for treating or preventing disease in human beings;

or:

> any substance or combination of substances which may be used in or administered to human beings either with a view to restoring, correcting or modifying physiological functions by exerting a pharmacological, immunological or metabolic action, or to making a medical diagnosis."

Especially the difficulties in the clear interpretation of "physiological function" in contrast to "pharmacological action" make the definition fuzzy and nonspecific for many products. There are already products on the market generally recognized as foodstuffs which may also have therapeutic effects. But primarily all their characteristics have to be taken into account before the decision can be made to confer on them the status of medicinal products.

If medicinal claims are made on the basis of its "traditional use" as defined in Directive 2004/24/EC on Traditional Herbal Medicinal Products, or the herb is considered medicinal by function, the product may be categorized as a "Traditional Herbal Medicinal Product", as long as the time-related criteria are met [3]. The European Medicines Agency (EMEA) is responsible for assessing both the safety and efficacy of herbal preparations when used as medicines.

9.2.2
Foods

The use of plant extracts in food is regulated under the General Food Law (178/2002/EC), which assigns primary legal responsibility for the safety of the products placed on the market to business operators [4]. This regulation establishes the European Food Safety Authority (EFSA) [5]. The regulation applies to all food stuffs and covers also food with added functional properties (e.g., "Novel Food", "Functional Food", "Food Additives" and "Food Supplements").

Article 2 states that: "Food" means any substance or product, whether processed, partially processed or unprocessed, intended to be, or reasonably expected to be ingested by humans: excluding

a) feed;

b) live animals unless they are prepared for placing on the market for human consumption;

c) plants prior to harvesting;

d) medicinal products within the meaning of Directives 2004/27/EC [2];

e) cosmetics within the meaning of Directive 76/768/EC [6];

f) tobacco and tobacco products within the meaning of Directive 89/622/EC [7];

g) narcotic or psychotropic substances within the meaning of the United Nations Single Convention on Narcotic Drugs, 1961, and the United Nations Convention on Psychotropic Substances, 1971 [8];

h) residues and contaminants.

This definition excludes medicinal products completely. But some botanicals are considered as traditional herbal medicinal plants and are used both in medicinal products and in food supplements. That means that the final decision to judge whether medicinal law or food law applies to a specific product will have to be sought in medicinal law.

If a natural product is categorized as food or food ingredient according to Article 2 of Regulation 178/2002/EC [4], it must be characterized and further defined according to the following categories of foods:

- Novel food (Directive 258/1997/EC) [9]

- Foods for particular nutritional use (PARNUTS), covering dietetic foods (Directive 89/398/EC) [10]

- Food additives (Directive 89/107/EC) [11]

- Food supplements (Directive 2002/46/EC) [12]

- Flavorings (Directive 91/71/EC completing Directive 88/388/EC) [13]. The Directive applies to flavorings used to impart odor and/or taste to food. The

Member States must take the necessary measures to ensure that flavorings cannot be marketed or used unless they comply with the conditions laid down in the Directive, such as the purity criteria and restrictions on dangerous or undesirable constituents or substances.

Interestingly "functional foods", the issue most discussed in marketing plant extracts in foods, are currently not regulated as a distinct category in the EU. Therefore marketing of plant extracts with health benefits in foods still causes problems, as there is no clear definition on this term, as for example in Japan. In general food products with specific medical or physiological benefits must comply with all relevant food legislation with the respect to the composition. Especially with labeling and claims, the Regulation on Nutrition and Health Claims (Regulation 1924/2006/EC) is important [14].

9.3
Application of Plant Extracts in Flavorings, Functional Foods, Novel Foods and Food Supplements

9.3.1
Flavorings

Flavorings are substances used to impart taste and/or smell to food. Food manufacturers have been using flavoring substances for many years in a wide variety of foods, from confectionary and soft drinks to cereals, cakes and yoghurts. EU legislation defines different types of flavorings, such as natural, natural-identical (chemically identical to natural substances but obtained by chemical processes) and artificial flavorings, as well as flavorings of plant or animal origin and smoke.

Member States have taken different regulatory approaches in the past, mainly opting not to regulate the use of flavorings. Because of questions regarding the safety of flavorings there was a need for regulation in the EU. The regulatory framework is now harmonized through EU legislation on food flavorings. (Directive 91/71/EC completing Directive 88/388/EC on the approximation of the laws of the Member States relating to flavorings for use in foodstuffs and to source materials for their production. Regulation 1334/2008/EC of 16 December 2008 on flavorings and certain food ingredients with flavoring properties for use in and on foods and Regulation 1331/2008/EC of 16 December 2008 establishing a common authorization procedure for food additives, food enzymes and food flavorings) [13, 15, 16].

The Regulation 2232/96/EC set out the main rules on the use of flavorings in foods in the EU. Based on this legislation a procedure was launched to establish an EU positive list of flavorings which will govern the flavorings that may be added to foods. The positive list is to be set up after a comprehensive safety evaluation program has been completed. Member States have notified the Commission of around 2800 substances authorized at national level to be included in the program. Among them are many substances which occur naturally [5].

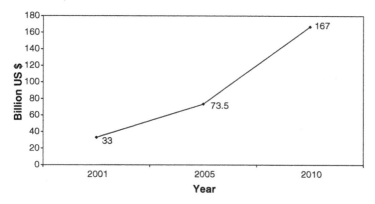

Figure 9.1 Development and forecast of the global functional food market [19–21].

9.3.2
Functional Foods

Functional foods are natural foods which may confer health benefits in addition to their basic nutritive value. They can and should be consumed as a part of the daily diet and have particular functions when ingested such as preventing specific diseases, slowing the aging process, control of physical and mental conditions and enhancement of the biological defense mechanism [17]. Functional foods contain ingredients like n-3 fatty acids for treating hypertriglyceridemia, plant sterols or stanols that may reduce serum cholesterol levels and probiotics which confer health benefits on the host [18]. These are only few examples of functional components in food. In recent years the use of functional foods has become a growth area for the food industry as shown in Figure 9.1. Thus, you find an increasing number of food on the market that contains plant extracts with special health effects and appropriate health and nutrition claims on their labels [22]. The benefits of plant extracts for the prevention and treatment of specific diseases and the regulations of food products with added plant extracts will be described in the following sections.

9.3.3
Regulation on Nutrition and Health Claims Made on Foods

There is an increasing number of food on the European market with labels, which bear nutrition and health claims. This kind of labeling is regulated by Regulation 1924/2006/EC of 20 December 2006 on nutrition and health claims made on foods [14]. The legislation on these claims ought to protect consumers by prohibiting the use of false and misleading information. Products with functional components have to be safe and adequately labeled. Further objectives of this Regulation are to encourage product innovation in the food industry (legal security), to promote fair trade and to ensure the functioning of the single European market.

The regulation applies to all nutrition and health claims made in commercial communication including labeling, advertising of food and promotional campaigns. It also applies to trademarks and brand names which may be interpreted as health or nutrition claims. Terms and conditions for labeling with nutrition and health claims:

- The physiological effects of the substance that is subject of the claim must be scientifically substantiated.
- A significant amount of the functional ingredient should be available in an adequate daily portion of the foodstuff.
- The regularly consumed amount must be considered.
- The bioavailability of the substance must be assured.

Food business operators should justify the use of a claim and ensure the scientific substantiation [14]. Examples of some basic criteria for the scientific substantiation of health claims on foods and food components are listed below:

- The food or food component should be well characterized and comply with existing legislation.
- Claims should be based on human intervention studies considering the target group, appropriate controls and adequate duration to demonstrate the intended effect.
- Scientific data which are generally accepted must be objective and systematic, strong, balanced, consistent, reproducible and not falsified, with appropriate statistical analysis.
- Studies should use scientifically validated biomarkers to measure the claimed enhancement of function or reduction of disease risk.
- Health claims must be evaluated by taking into account the total evidence.
- The assessment of significant scientific agreement should be based on critical interpretation of the data and on the application of scientific judgment [6].

A nutrition or health claim must not be made if it is inconsistent with generally accepted nutrition and health principles or if it leads to a one-sided dietary practice. The use of health claims in beverages containing more than 1.2% of alcohol by volume is generally prohibited. Consumers should be able to estimate the nutritional quality of functional food. According to the Directive 90/496/EC of 24 September 1990 on nutrition labeling for foodstuffs labels should bear nutritional information including the energy value and the quantities of carbohydrates, sugars, fats, proteins, sodium und fibers [23].

A new claim has to be authorized by the European Commission. Therefore manufacturers must submit an application to the national competent authority of a Member State. The decision on the use of the applied claim is based on the opinion of the EFSA. Furthermore the new claim must appear on the community list. The adoption of this community list of permitted claims and the conditions for the use of these claims is progressive through 2010 and 2011 [24]. According to technological and scientific developments this community list will be updated at regular intervals.

Further conditions for the use of claims are specific nutrient profiles for food. These profiles will be arranged in consideration of the quantities of fat, saturated fatty acids, trans-fatty acids, sugars and salt/sodium. The importance of these food products for feeding the population, particular risk groups and children will be taken into account. Nevertheless the nutritional value of the food and the content of nutrients with scientifically proved health effects have to be estimated. These nutrient profiles will be scientifically validated by the EFSA. They shall prohibit that food with inadequate nutrient profiles (e.g., high sugar content or high percentage of trans-fatty acids) gives the impression of being especially healthy because of a functional component which has been added (e.g., sweets enriched with vitamins). In the following section, definitions of the different claims (taken from Article 2 of the Regulation 1924/2006/EC [14]) and appropriate examples will be given.

9.3.3.1 Nutrition Claim
"Nutrition claim means any claim which states, suggests or implies that a food has particular beneficial nutritional properties due to:

a) the energy (calorific value) it
   - provides,
   - provides at a reduced or increased rate, or
   - does not provide, and/or
b) the nutrients or other substances it
   - contains,
   - contains in reduced or increased proportions, or
   - does not contain."

See nutrition claims and conditions applying to them in the annex of the Regulation 1924/2006/EC [14]. The application of nutrition claims will only be allowed if they are included in this annex and fulfill the appropriate specific conditions.

An example of a nutrition claim is as follows:

> "High fibre – A claim that a food is high in fibre, and any claim likely to have the same meaning for the consumer, may only be made where the product contains at least 6 g of fibre per 100 g or at least 3 g of fibre per 100 kcal."

9.3.3.2 Health Claim
"Health claim means any claim that states, suggests or implies that a relationship exists between a food category, a food or one of its constituents and health." Specific conditions and requirements for the use of health claims are described in

Chapter IV of Regulation 1924/2006/EC [14]. An example of an applied health claim that appears on the consolidated community list of Article 13 health claims (No: 1310) is:

"Food: Epigallo-catechin-3-gallate (EGCG)/Green tea extract, rich in EGCG

Health relationship: Cardiovascular health

Example of wording: Contributes to improved blood flow/in combination with physical activity helps to reduce blood pressure/contributes to maintain a healthy cardiovascular system" [25].

9.3.3.3 Reduction of Disease Risk Claim

"Reduction of disease risk claim means any health claim which states, suggests or implies that the consumption of a food category, a food or one of its constituents significantly reduces a risk factor in the development of a human disease" [14]. An example of a reduction of disease risk claim pursuant to Article 14 of Regulation 1924/2006/EC is the following:

"Plant stanol esters have been shown to lower/reduce blood cholesterol. Blood cholesterol lowering may reduce the risk of coronary heart disease" [26].

9.3.4
Regulation Concerning Novel Foods and Novel Food Ingredients

It must be proved whether or not new food products containing plant extracts belong to the category of novel foods. Novel foods are defined and regulated by the Regulation 258/97/EC of 27 January 1997 concerning novel foods and novel food ingredients. This Regulation comes into effect when placing on the market within the community those foods or food ingredients which were not used for human consumption in a considerable range before 15 May 1997.

The following categories of foods fall in the scope of the Regulation 258/97/EC:

- Foods and food ingredients which consist of or are isolated from microorganism, algae or fungi.

- Foods and food ingredients consisting of or are isolated from plants and ingredients isolated from animals except those obtained by traditional propagating or breeding practices and having history of safe food use.

- Foods and food ingredients which have a new or intentionally modified primary molecular structure.

- Foods and food ingredients which have been subject to a new production process not currently used, where that process is resulting in significant changes in the composition or structure of the foods or food ingredients which affect their nutritional value, metabolism or level of undesirable substances [9].

This Regulation does not apply to food additives, flavorings for use in foodstuffs or extraction solvents used in the production of foodstuffs. Novel foods and food ingredients must not present a danger for the consumer and mislead him or her. Furthermore they must not be nutritionally disadvantageous for the consumer. Therefore these products have to be analyzed in order to protect public health. Before novel food and novel food ingredients are placed on the European market they have to be subject to a single safety assessment through a community procedure in order to get the approval. Manufacturers must submit a request to the Member State in which the product is to be placed on the market for the first time and at the same time they shall send a copy of the request to the European Commission [9].

Predominantly juices of the fruits of *Morinda citrifolia* known as "noni-juice" and various foodstuffs with added phytosterols/phytostanols currently belong to the approved novel food group. Examples for some novel food notifications in the recent years are:

- Noni-juice or juice of the fruits of *M. citrifolia* (2003)
- Astaxanthin-rich carotenoid oleoresin extracted from *Haematococcus pluvialis* (2004)
- DHA (docohexanoid-acid)-rich micro-algal oil (2003)
- Milk type products and soya drinks with added phytosterols/phytostanols (2004)
- Potato protein (2007)
- Argan oil (2007)
- Extract from *Ajuga reptans* (2008)
- Leaves of *M. citrifolia* (2008) [27].

9.3.5
Regulation Concerning Food Supplements

In addition to functional foods and novel foods there are products on the market containing concentrated sources of nutrients. These are called food supplements or dietary supplements. They are intended to complete the normal diet by ingestion in the form of capsules, pastilles, tablets, pills, sachets of powder, ampoules of liquids and other similar forms. They are designed to be taken in particular amounts of small unit quantities. Dietary supplements have been regulated since 2002 in the Directive 2002/46/EC of 10 June 2002 on the approximation of the laws of the Member States relating to food supplements.

Under normal circumstances a varied and balanced diet could provide the necessary nutrients for normal development and maintenance of a healthy life. Surveys show that this situation is not being achieved for all nutrients in the feeding of the population across the community. Due to particular lifestyles or diseases consumers choose dietary supplements to complete their personal diets.

Food supplements that will be put on the market should be proven to be safe in quantity and quality and bear adequate labeling. Legislation should ensure high protection for consumers. At present the Directive 2002/46/EC only applies to vitamins and minerals while there is a wide range of food supplements on the market containing amino acids, essential fatty acids, fibers and various plants and herbal extracts. Only vitamins and minerals listed in Annex I and vitamin and mineral substances listed in Annex II of the Directive 2002/46/EC may be used in manufacturing food supplements.

The scope of the Directive will be extended to include various ingredients when appropriate scientific data of ingredients with nutritional or physiological effects such as plant extracts become available. Food supplements with botanical ingredients are already covered by national legislation in Austria, Belgium, the Czech Republic, Finland, France, Greece, Italy and Slovenia [1, 12].

9.3.6
The Health Benefits of Food Products Containing Plant Extracts

Mainly foods such as beverages, sweeties, bakery products, health food and food supplements belong to the products, which are enriched with plant extracts. The goal of enriching these foods with plant extracts is to provide health benefits. The plant extracts are taken from different basic raw materials. Some of the most common plants with potential medical benefits are black and green tea, guarana, gingko biloba, ginseng, garlic, grape seed and soy [22, 28, 29]. From these plants secondary plant substances such as polyphenols, isoflavones, carotenoids, and others, are extracted. These active compounds act in the human body in a range of capacities, especially as antioxidants, cholesterol fighters or immune boosters. Secondary plant substances are found both in seeds and in the leaves of the plants. These active ingredients can be ingested in form of dietary supplements or added to foods [29].

9.3.6.1 Mode of Function Plant Extracts as Antioxidants
The active compounds of some plant extracts function partially as antioxidants and thereby support the inactivation of free radicals. Free radicals are atoms, molecules or ions with an unpaired electron (e^-) which are usually highly reactive. They may cause a chain of chemical reactions that damage cells. The human body possesses some defense mechanisms to stop these reactions. Antioxidants in foods support these processes and therefore reduce the disease risk, particularly the development of cancer [28].

9.3.7
Examples of Applications of Plant Extracts in Functional Foods and Dietary Supplements

The following section describes a few plant extracts, their application in food and dietary supplements as well as their health benefits. Table 9.1 gives an overview

Table 9.1 Plant extracts and the appropriate functional components with beneficial or health protective properties [28, 30–33].

Plant extract	Functional components	Beneficial or health protective properties
Aged garlic extract	S-allylcysteine, allyl sulfides	Reduces total serum cholesterol level, reduces cancer risk, boosts immune function
Black tea extract	Proanthocyanidins, bisflavonols, theaflavins and thearubigin	Antioxidant
Fibers	Important soluble fibers among others are pectins, celluloses, beta-glucans, inulin, oligofructose, resistant starch and psyllium	Reduce serum lipid levels, increase fecal bile acid excretion, slow absorption of dietary sugars, decrease risk of cancer and coronary artery disease
Gingko biloba	*Gingko biloba* leaf extract	Stabilizes or improves dementia, anti-inflammatory, antioxidant
Ginger	6-gingerol	Antioxidant
Ginseng	Ginsenosides	Antioxidant, anti-inflammatory
Grape seed extract	Range of proanthocyanidins	Antioxidant
Green tea extract	Epigallocatechingallate (EGCG) and related catechins, theogallin	Antioxidant
Soy isoflavones	Predominantly genistein, daidzein	Phytoestrogen – alternative to hormone replacement therapy
Soy protein	Not specified so far – probably soy protein including isoflavones	Lowers levels of total serum cholesterol and low-density lipoprotein (LDL-cholesterol), decreases risk of coronary artery disease

of various plant extracts and the appropriate functional components with beneficial or health protective properties.

9.3.7.1 Black and Green Tea

Tea (*Camellia sinensis*) is a common beverage around the world, particularly in Asia, and is known for its functional and healing effects. Because of this, tea has become popular as an extract in functional foods and food supplements in Asian countries as well as in recent years in the Western world [17, 34]. Tea contains many physiologically effective phenolic compounds. The most important representatives of these so-called flavonoids are catechin, epicatechin, gallocatechin, epigallocatechin, epigallocatechingallat and epicatechingallate. These compounds have, among others, cardiovascular-protective and anticarcinogenic properties which can be traced back to their antioxidant capacities [29, 35]. In the USA statements are printed on the teabag packages about the content of flavonoids ("on-pack claims"), for example: "You might have heard about antioxidants in fruits and vegetables. But did you know that ... Tea contains 175 mg of antioxidants per serving? ... Green Tea, which comes from the same plant, contains 135 mg of antioxidants per serving. Research suggests that antioxidants can help neutralize "free radical" molecules that can damage cells" [29]. Further potentially physiological side effects of tea are:

- Antidiabetic side effects
- Decrease of elevated blood lipids
- Polyphenols in tea inhibit streptococcus enzymes and reduce the formation of plaque, therefore functioning against caries.

In several studies it is reported, that green tea and epigallocatechin gallate (EGCG) have positive effects on body weight loss and the reduction of adipose tissue. Therefore a daily consumption of tea or beverages enriched with tea extract may be useful in weight management and the prevention of lifestyle-related diseases [36, 37]. Tea extracts are put in "ready to drink" beverages in different flavors. A very popular refreshment is "Ice-Tea". Although a minimum amount of tea extract of $1 g L^{-1}$ should be found in beverages, the amount of polyphenols varies greatly due to the technological changes of the extracts. For example, substantial amounts of flavonoids are removed by using ethyl acetate when producing decaffeinated beverages for children. The selection of the solvents used is of great importance for the concentration of the functional element in the food. Tea extracts are also used in confectionary. Due to the amount of flavonoids added, one can then advertise the health benefits [29].

9.3.7.2 Gingko Biloba

The gingko tree belongs to the oldest living tree species and the seeds are used as food in China and also as medication for treating various diseases. Gingko leaves have become interesting in the last decades as medication for treating circulatory disorders and for improving cerebral circulation, which may enhance mental function. The leaves of *Gingko biloba* also contain chemical compounds (flavonoids and terpenoids) which are supposed to have antioxidant properties. Thus, these antioxidants can neutralize free radicals and may reduce some of the damages they cause. Gingko leaves are used in form of concentrated standardized extract in dietary supplements, refreshments and herbal teas and have become very popular in the health food market in Europe and in the USA [17, 22, 30].

9.3.7.3 Soy Protein

Soy is an interesting foodstuff in many aspects. Soy products are to be found in several "meat analog" products, such as soy sausages, soy cheese and yogurts which are intended to replace animal-based food. Vegetarians and people with healthy lifestyle also consume soy products like tofu, miso, tempeh and soymilk instead of meat, because soybeans and most soy products are good sources of high quality protein. Soy protein contains all essential amino acids which are necessary for the human diet. It is a good alternative to cow`s milk for those who have cow's milk protein allergy or lactose intolerance. Further benefits of soy products compared with animal-based food are that they are low in saturated fats and cholesterol-free [31].

Improvement of cardiovascular health is another potential advantage [38]. A meta-analysis reported that an average intake of 47 g soy protein per day, without changing dietary saturated fats or cholesterol, reduces serum total cholesterol by 9% and LDL cholesterol levels by 13%. The effect may be caused by the protein itself or by plant substances such as isoflavones or saponins [32]. Since 1999 the Food and Drug Administration (FDA) of the USA accepts health claims on soy products, because soy protein is generally recognized as safe. Also, there are no restrictions for adding soy protein to food. To qualify soy products for health claims on their labels they must have a low fat, sodium, saturated fat and cholesterol content and 6.25 grams soy protein per serving. The appropriate claim could be as follows:

> "Diets low in saturated fat and cholesterol that include 25 grams of soy protein a day may reduce the risk of heart disease. One serving of (name of food) provides x grams of soy protein" [31].

Specific components of soy are isoflavones which are also found in low amounts in legumes. Isoflavones are phytoestrogens which are used in form of isoflavone extracts in dietary supplements for the improvement of menopausal symptoms as an alternative to hormone replacement therapy. Studies report a reduction of hot flushes in peri- and postmenopausal women. Scientists say that isolated isoflavones could have bad effects if consumed in large amounts for a long time. The safe range of intake is 35–55 milligrams a day (Japanese consume averagely one serving of soy food per day, equivalent to 25–50 mg isoflavones).

Some clinical studies show that the consumption of high amounts of soy prevents the risk of breast cancer in women and some studies even show the opposite. The influence of isoflavones on risk of breast cancer seems to depend on the age at which it was consumed. Young women may reduce the risk of breast cancer by consuming soy products while postmenopausal woman may increase their cancer risk by consuming high amounts of soy [31, 39, 40].

9.4
Requirements Concerning Quality and Safety Issues

The use of natural products in food is governed by regulatory laws of the country or international standards. Internationally the Joint FAO/WHO Expert Committee for Food Additives (JECFA) evaluate the available information of potential risks and benefits to reach informed decisions on food safety of substances and extracts added to food. In the EU the use of botanicals and botanical preparations in food

Table 9.2 Examples for biochemicals extracted from foods and by products used in foods.

Source	Product	Application
Soya	Lecithin	Emulsifier
Egg	Lysozyme	Preservative
Black carrots	Anthocyans	Colorant
Vanilla	Vanillin	Aroma
Beetroot	Betaine	Colorant
Stevia rebaudiana	Steviosides, rebaudiosides	Intense sweetener
Rosemary	Rosmarinic acid	Antioxidant
Fish, algae, vegetable oil	Omega 3 fatty acids	Food supplement
Papaya	Papaine	Protease for meat tenderizing

has to be in compliance with the existing General Food Law (178/2002/EC) [4]. This would include maximum permissible levels of chemical and biological contaminants (e.g., pesticides, mycotoxins, heavy metals and food-borne pathogens) according to the Regulations 1881/2006/EC [41] and 396/2005/EC [42], modalities for ensuring quality, and application of good hygienic practice, including HACCP methodologies (Regulation 852/2004/EC) [43]. However, the regulations do not provide guidance on how the safety of these products should be assessed. But there are some general concerns with respect to botanicals and botanical preparations mainly relating to quality and safety issues.

The chemical structure of these food ingredients is as different as the natural chemistry could be. All classes of active structures like isolated phytochemicals, protein fractions and hydrolysates, enzymes, essences from herbs and spices and other plant extracts, fractionated fats, fibers, starch, sugar, vitamins and colorants could be isolated from plants and used in foods (Table 9.2). The purification of these materials often requires sophisticated separation techniques.

Contamination (both chemical and microbiological) is still a documented problem. It has been associated, for example, with botanical products originating from Asia. Deaths through poisoning following consumption of such products have been reported in Europe and the USA arising from contamination with heavy metals, synthetic drugs and other undesirable substances [44]. Misidentification of plants harvested from the wild is also a continuing problem. When sales of traditional herbal products remained restricted to particular ethnic groups, which are familiar with the products and able to exert a local control on quality, problems were more rarely encountered. The growing volume of sales in the European Union with products obtained from suppliers based in Asia or elsewhere and the move towards widespread outlets for the products of traditional medicine call for more formal pre-marketing assessment and more stringent controls than the occasional random checks and analyzes often carried out by individual national or local authorities on what is already out in the market [45].

A very large literature exists on toxic substances naturally-occurring in plants. Of particular concern are plant products consumed in concentrated forms.

Concerns can relate directly to the toxicological profile of the active agent(s). It has been the case with kava-kava (*Piper methysticum*), its withdrawal from the market being advised recently because of the association of its consumption with acute liver failure [46]. In June 2001, the US FDA, following action by the UK and Australia, advised manufacturers of dietary supplements to avoid the use of the various types of comfrey (*Symphytum* spp.) because of the health concerns associated with the presence of pyrrolizidine alkaloids. Moreover, since 1993, cases of nephrotoxicity and carcinogenicity have been reported in Belgium, France and United Kingdom as a result of inadvertent exposure to *Aristolochia* species in unlicensed herbal medicines [47].

Because of the loophole in the European legislation scientific societies like the German Chemical Society (GDCh) or the International Life Sciences Institute (ILSI Europe) drafted proposals for quality standardization of botanical preparations. The proposals claim that it is essential that the plant products should be well identified and characterized [48, 49].

There are three aspects to the production of a consistent material from natural plants:

1. Identification of starting material
2. Process applied to starting material
3. Standardization and nature of final product

On the European level the EFSA published on 9 September 2009, an updated draft of a guidance document on the safety assessment of botanicals and botanical preparations [50]. This work was initiated towards the end of 2005 after the members of EFSA's Advisory Forum underlined the need for science-based guidance for assessing the safety of botanicals.

The data required in this guidance aims at identifying and assessing the possible hazards associated with botanicals and botanical preparations. These data are of: (i) technical; (ii) exposure and (iii) toxicological nature. The lists below are meant to give guidance on possible data requirements for any operator or organization in charge of assessing the safety of botanical ingredients. They have been made as exhaustive as possible and should be adapted on a case-by-case basis, depending on the nature of the botanical or botanical preparation. This implies that not all the information listed would be needed in all cases and that the amount of information available for a given botanical or botanical preparation may in some cases be sufficient without further testing.

9.4.1
Technical Data

9.4.1.1 Identity and Nature of the Source Material
- Identity, that is scientific (Latin) name (botanical family, genus, species, variety with author's name, chemotype if applicable).
- Common names.
- Part used (e.g., root, leaf, seed...).

- Geographical origin (continent, country, region).
- Growth and harvesting conditions (wild or cultivated; cultivation practices, time of harvest in relation to both season and stage of the plant growth).
- In the case of cultivated plants, the origin of the seed or propagules.

9.4.1.2 Manufacturing Process

The following information is considered necessary for assessing the safety of botanicals and botanical preparations:

i) Information on the method(s) of manufacture (e.g., the process by which the raw material is converted into a preparation, such as extraction or other procedure(s), and plant extract ratio)

ii) Information on substances entering the manufacturing process, e.g., identity of the extraction solvent, reagents, special precautions (light and temperature).

iii) Standardization criteria (e.g., see European Pharmacopoeia).

9.4.1.3 Chemical Composition

Data on the chemical composition of the botanical ingredient should be provided with emphasis on compounds of relevance for the safety assessment.

9.4.1.4 Specifications

Specifications of the botanicals or botanical preparations are required. They may be based on nutritional or biologically active components or, when these are not known, on selected chemical markers. Limits for or absence of specific undesirable/toxic substances should be specified. The proposed specifications should be modeled on recent European or other internationally accepted specifications (e.g., pharmacopoeia or the guidelines of the European Medicines Agency (EMEA) Committee on Herbal Medicinal Products (HMPC). Where the proposed specifications differ from internationally recognized specifications, the latter specifications should be set out alongside the proposed new specifications, and any differences pointed out. Validated and well-established methods should be preferably used for the analysis of compounds considered in specifications.

9.4.1.5 Stability of the Botanical or Botanical Preparation used as Ingredient in Food Supplement

The stability of the botanical ingredient should be demonstrated over the shelf-life time. Any information concerning possible degradation should also be provided.

9.4.1.6 Proposed Uses and Use Levels

Information should be provided on the intended uses and recommended intakes for each botanical ingredient.

9.4.1.7 Information on Existing Assessments

Information on any existing assessments by national competent authorities or other bodies should be provided.

9.4.2
Additionally Data on Exposure and Toxicological Exposure: Extent and Time

Data and information should be provided on:

i) Anticipated human exposure to the botanical ingredient, including amount (e.g., maximum and average daily intake or exposure), frequency and duration. It is important to characterize as much as possible the expected human exposure to the botanical ingredient according to the recommended modalities of use in terms of extent and time.

ii) Possibility of additional/combined human exposure to the botanical or botanical preparation through different categories of food, food supplements and/ or medicinal products that can be consumed together.

iii) Modality of use of the ingredient.

iv) Information on historical (food, food supplement and medicinal) use of the ingredient in human population groups in relation to the modalities of use and resulting exposure levels if known. Data derived from use outside of the European Union should also be considered. All data should be representative of the ingredient to be used for the European market. In this context, food use includes, in principle, the consumption of raw and cooked vegetables, spices, flavors, food supplements already in use for a long time and any other related food items.

Estimates of average exposure ranges associated with the use of the botanical ingredient in the EU Member States should also be provided. Uncertainties associated with the food consumption data considered and anticipated exposure ranges should be clearly described [17].

9.4.3
Toxicological Data

Studies on toxicity and toxicokinetics including metabolism of botanicals and botanical preparations should be conducted using internationally agreed protocols. Test methods described by OECD or in European Commission Directives 87/432/ EC and 67/548/EC–Annex 5 are recommended [51, 52]. It is advisable to ensure that the most up-to-date version of any test guideline is followed. Use of any methods differing from internationally agreed protocols should be justified. Protocols for special studies differing from standard tests should be developed on a case-by-case basis [51].

9.5
Conclusions

Although the application of plant extracts in food is dealt by a multitude of regulations, still the red thread of an unambiguous definition, labeling and standardization is missed in the EU. The demarcation dispute of medical or food status is still open and the approval is up to the member State in the EU. Companies producing and marketing natural food products have to assess their regulatory status individually and nationally specific. Since 2004 the EFSA has made great efforts to facilitate the health food market of plant extracts with improved regulations to optimize safety issues and consumer information.

A new regulation for Nutrition and Health claims has been developed with strong requirements on scientific evidence. Recently in June 2008 the EFSA has published a toolkit composed of a framework, compendium and checklist to prioritize and handle the safety assessment of natural plant products more efficiently. These guidelines are great operation instructions for food business operators to evaluate critical pre-marketing questions and preparing approval situation in a professional way. The end-consumers benefit from a higher safety level of plant extracts because of improved standardization and scientific provement.

References

1 Bagchi, D. (ed.) (2008) *Nutraceutical and Functional Food Regulations in the United States and around the World*, Elsevier, Burlington, USA, pp. 199–219.
2 Directive 2004/27/EC of the European Parliament and of the Council of 31st March, 2004 on the Community code relating to medicinal products for human use amending Directive 2001/83/EC, *Official Journal of the European Union*, L 136/34, 30.04.2004.
3 Directive 2004/24/EC of the European Parliament and of the Council of 31st March 2004 amending, as regards traditional herbal medicinal products, Directive 2001/83/EC on the Community code relating to medicinal products for human use, *Official Journal of the European Union*, L 136/85, 30.04.2004 on the Community code relating to medicinal products for human use, *Official Journal of the European Union*, L 311, 28.11.2001.
4 Regulation 178/2002/EC of the European Parliament and of the Council of 28 January 2002 laying down the general principles and requirements of food law, establishing the European Food Safety Authority and laying down procedures in matters of food safety, *Official Journal of the European Union* L 31, 1.2.2002.
5 European Food Safety Authority (EFSA) (2009) www.efsa.europa.eu (accessed at 31.08.2009).
6 Council Directive 76/768/EC of 27 July 1976 on the approximation of the laws of the Member States relating to cosmetic products, *Official Journal of the European Union* L 53, 25.2.1977.
7 Council Directive 89/622/EC of 13 November 1989 on the approximation of the laws, regulations and administrative provisions of the Member States concerning the labelling of tobacco products, *Official Journal of the European Union* L 359, 8.12.1989.
8 Final Act of the United Nations Conference to consider amendments to the single Convention on Narcotic Drugs, 1961 as amended by the 1972 Protocol amending the Single Convention on Narcotic Drugs, 1961.

9. Regulation 258/97/EC of the European Parliament and of the Council of 27 January 1997 concerning novel foods and novel food ingredients, *Official Journal of the European Union* L43, 18.04.2004.
10. Council Directive 89/398/EC of 3 May 1989 on the approximation of the laws of the Member States relating to foodstuffs intended for particular nutritional uses, *Official Journal of the European Union* L 275, 5.10.1990.
11. Council Directive 89/107/EC of 21 December 1988 on the approximation of the laws of the Member States concerning food additives authorized for use in foodstuffs intended for human consumption, *Official Journal of the European Union* L 40, 11.2.1989.
12. Directive 2002/46/EC of the European Parliament and of the Council of 10 June 2002 on the approximation of the laws of the Member States relating to food supplements (Text with EEA relevance), *Official Journal of the European Union* L 183, 12.7.2002.
13. Commission Directive 91/71/EC of 16 January 1991 completing Council Directive 88/388/EEC on the approximation of the laws of the Member States relating to flavorings for use in foodstuffs and to source materials for their production, *Official Journal of the European Union* L 42, 15.2.1991.
14. Regulation 1924/2006/EC of the European Parliament and of the Council of 20 December 2006 on nutrition and health claims made on foods, *Official Journal of the European Union* L404, 18.01.2007.
15. Regulation 1334/2008/EC of the European Parliament and of the Council of 16 December 2008 on flavorings and certain food ingredients with flavoring properties for use in and on foods and amending Council Regulation 1601/91/EC, Regulations 2232/96/EC and 110/2008/EC and Directive 13/2000/EC (Text with EEA relevance), *Official Journal of the European Union* L 354, 31.12.2008.
16. Regulation 1331/2008/EC of the European Parliament and of the Council of 16 December 2008 establishing a common authorisation procedure for food additives, food enzymes and food flavorings (Text with EEA relevance). *Official Journal of the European Union* L 354, 31.12.2008.
17. Goldberg, I. (1999) *Functional Foods: Designer Foods, Pharmafoods, Nutraceuticals*, Chapman & Hall, New York, USA, p. 393.
18. Heimburger, D.C. and Ard, J.D. (2006) *Handbook of Clinical Nutrition*, 4th edn, Mosby Elsevier, Philadelphia, USA, p. 219.
19. Menrad, K. (2003) *J. Food Eng.*, **6** (2–3), 181.
20. ReportLinker, (2009) Global market review of functional food–forecasts to 2012, Summary, http://www.reportlinker.com/p038872/Global-market-review-of-functional-foods-forecasts-to-2012.html (accessed 8 June 2009).
21. Business Wire, (2005) Research and Markets: Current Global Functional Foods Market Estimated to Grow to US$167bn by 2010, http://findarticles.com/p/articles/mi_m0EIN/is_2005_Feb_28/ai_n11832245/ (accessed 18 June 2009).
22. Meyer, A.H. (2009) Marktübersicht, in *Praxishandbuch Functional Food* (eds H.F. Ebersdobler and A.H. Meyer), Behrs Verlag, Hamburg, Germany, p. 39. Aktualisierungs-Lieferung 04/2009, Ordner 4, Teil III, Kap. 8.
23. Council Directive 90/496/EEC of 24 September 1990 on nutrition labelling for foodstuffs, *Official Journal of the European Union* L 276, 6.10.1990.
24. European Commission (2010), Regulation (EC) No 1924/2006 nutrition and health claims on foods, http://ec.europa.eu/food/food/labelling nutrition/claims/docs/doc1_nat_impl_reg_20100309.pdf, (accessed at 1.12.2010)
25. European Food Safety Authority (EFSA) (2009) Article 13, database with the list of health claims, http://www.efsa.europa.eu/EFSA/efsa_locale-1178620753812_article13.htm (accessed 5 May 2009).
26. European Food Safety Authority (EFSA) (2009) Plant stanol esters and blood cholesterol–scientific substantiation. http://www.efsa.europa.eu/EFSA/efsa_

locale-1178620753812_121 (accessed 5 May 2009).
27 European Commission (2009) Notifications pursuant to Article 5 of Regulation (EC) No 258/97 of the European Parliament and of the Council. http://ec.europa.eu/food/food/biotechnology/novelfood/notif_list_en.pdf (accessed 9 April 2009).
28 Watson, R.R. (ed.) (2003) *Functional Foods & Nutraceuticals in Cancer Prevention*, Iowa State Press, Iowa City, USA.
29 Engelhardt, U. (2009) Flavonoide (Polyphenole), in *Praxishandbuch Functional Food* (eds H.F. Ebersdobler and A.H. Meyer), Behr's Verlag, Hamburg, Germany, p. 39. Aktualisierungs-Lieferung 04/2009, Ordner 1, Kap. 6.1.
30 Bridi, R., Crossetti, F.P., Steffen, V.M. and Henriques, A.T. (2001) *Phytother. Res.*, **15** (5), 449.
31 Henkel, J. (2009) Soy, Health Claims for Soy Protein, Questions About Other Components, http://www.fda.gov/Fdac/features/2000/300_soy.html#health (accessed 25 May 2009).
32 Anderson, J.W., Johnstone, B.M. and Cook-Newell, M.E. (1995) *N. Engl. J. Med.*, **333**, 276.
33 Jacobasch, G. and Dongowsky, G. (2009) Ballaststoffe mit spezifischen Wirkungen, in *Praxishandbuch Functional Food* (eds H.F. Ebersdobler and A.H. Meyer), Behr's Verlag, Hamburg, Germany, p. 39. Aktualisierungs-Lieferung 04/2009, Ordner 1, Kap. 2.1.
34 Wichtl, M. (1994) *Herbal Drugs and Phytopharmaceuticals, A Handbook for Practice on A Scientific Basis* (ed. N.G. Bisset), Medpharm Scientific Publishers, Stuttgart, Germany, p. 490.
35 Mennen, L., et al. (2004) *J. Nutr.*, **134**, 923.
36 Nagao, T., et al. (2005) *Am. J. Clin. Nutr.*, **81**, 122.
37 Sayama, K., Lin, S., Zheng, G. and Oguni, I. (2000) *In Vivo*, 14 (4), 481.
38 Matthew, B.C. and Daniel, R.S. (2008) *J. Clin. Biochem. Nutr.*, **43**, 129.
39 Hickman, D. (2003) Soy food and breast cancer, in *Functional Foods &*
Nutraceuticals in Cancer Prevention (ed. R.R. Watson), Iowa State Press, Iowa City, USA, pp. 173–177.
40 Brown, J.L. (2001) Soy Protein and Soy Isoflavones, The Pennsylvania State University, Pennsylvania. http://pubs.cas.psu.edu/freepubs/pdfs/uk094.pdf (accessed 25 May 2009).
41 Commission Regulation (EC) No 1881/2006 of 19 December 2006 setting maximum levels for certain contaminants in foodstuffs (Text with EEA relevance), *Official Journal of the European Union* L 364, 20.12.2006.
42 Regulation 396/2005/EC of the European Parliament and of the Council of 23 February 2005 on maximum residue levels of pesticides in or on food and feed of plant and animal origin and amending Council Directive 91/414/EECText with EEA relevance. *Official Journal of the European Union* L 70, 16.3.2005.
43 Corrigendum to Regulation 852/2004/EC of the European Parliament and of the Council of 29 April 2004 on the hygiene of foodstuffs (*Official Journal of the European Union* L 139, 30.4.2004. Corrected version in *Official Journal of the European Union* L 226, 25.6.2004), *Official Journal of the European Union* L 204, 4.8.2007.
44 Ernst, E. (2002) *J. Int. Med.*, **252**, 107.
45 European Food Safety Authority (EFSA) (2004) EFSA/SC/26 Final. Discussion Paper on Botanicals and Botanical Preparations widely used as food supplements and related products: Coherent and Comprehensive Risk Assessment and Consumer Information Approaches (adopted by written procedure on 23 June 2004).
46 Kraft, M., et al. (2001) *Dtsch. Med. Wochenschr.*, **126**, 970.
47 European Medicines Agency (EMEA) (2000) Position paper on the risks associated with the use of herbal products containing Aristolochia species. EMEA/HMPWP/23/00.
48 Working Group Nutritional Issues of the German Society of Food Chemistry (2005) *Evaluation Guidelines for Plant Extracts in Food: Example Secondary Plant*

Products, Lebensmittelchemie, vol. 59, p. 107.

49 Schiltera, B., *et al.* (2003) *Food Chem. Toxicol.*, **41**, 1625.

50 European Food Safety Authority (2009) (Question No EFSA-Q-2009-00668) Guidance document of the Scientific Committee: Safety assessment of botanicals* and botanical preparations** intended for use as ingredients in food supplements. Published: 9 September 2009, EFSA, Parma, Italy.

51 Council Directive 87/432/EC of 3 August 1987 on the eighth adaptation to technical progress of Directive 67/548/EEC on the approximation of laws, regulations and administrative provisions relating to the classification, packaging and labelling of dangerous substances, *Official Journal of the European Union* L 38, 11.2.1988.

52 Council Directive 67/548/EC of 27 June 1967 on the approximation of laws, regulations and administrative provisions relating to the classification, packaging and labelling of dangerous substances, *Official Journal of the European Union* L 196, 16.8.1967.

Index

a

ABS, *see* alkylbenzenesulfonate
active pharmaceutical ingredient, *see* API
activity coefficient   28ff., 50f.
agar   227ff.
– industrial manufacture   229
agaropectin   228
agarose   228
alginate   227, 232ff.
alginic acid   233
– production of   236
alkylbenzenesulfonate   56, 62, 65
[Amim]   223
amino acid   29, 33, 44, 49
anhydrous thebaine alkaloid   265
– amounts of   266
antibody separation   148
antioxidant   278ff., 282
API   3, 248, 260f., 263f., 266
– blockbusters   260
– in plants   5
– list of   260
– recovery   7
application
– cloud point extraction   78
– micellar extraction   78
– reverse micelles   81
artemisinin
– Solubility data   63
– Structure   62
ATA, *see* anhydrous thebaine alkaloid

b

baobab   257ff.
batch   203
– -cycle   204
Berthelot–Lorentz rule   31
beta-carotene   38
– binary parameter   38
– PC-SAFT (lines)   39
– pure-component parameters   38
binary interaction parameter   31, 35, 50
biotransformation   221
[Bmim]   56, 59ff., 64, 223
BMOEA   56, 58, 62f.
botanicals   269ff., 281ff.
brand names   274
1-butanol   34, 40, 42

c

carotene   239f.
carrageenan   227, 230ff., 238
– alkali-modified flour (AMF)   231
– ι-carrageenan   230f.
– κ-carrageenan   230
– λ-carrageenan   230
– industrial manufacture   232
– semirefined (SRC)   231
cellulose   222
– crystalline   223
cephalotaxine   264
chlorophyll   239ff.
cholesterol   273, 276, 278f., 281
cholesterol oxidase   77
chromatogram   133ff., 147f.
– of recycling chromatography   142
– of simulated moving bed fractionation   142
– reduced   140
– reduction of   137
chromatographic separation   140
chromatography   11
– carrousel absorbers   23
– continuous   19
– – annular   23
– – SMB   20f.
– – TMB   20
– gel   19

– HPLC   126
– ion pair   18
– liquid–liquid (LLC)   18
– liquid–solid   17
– normal phase NP   18, 125
– preparative   136f.
– reversed phase RP   18, 125
– size exclusion   19
– thin layer (TLC)   133
claim   274
– Disease Risk   276
– Nutrition   275
cloud point   72f.
co-current   195
community list   274ff.
concentrate of poppy straw   264
contamination   269, 282
Counter-current   16, 186, 205
CP   91ff.
– critical points of some fluid   93
CPS concentrate of poppy straw   264f.
– amounts of   266
critical point, see CP
cross-current   205
crystals   63, 68f.
crystals
– liquid crystals   68
– turbid liquid anisotropic   68
CSEP®   23
curacin A   3

d

DAD, see diode-array detection
10-deacetylbaccatin   169f., 262
decorticating   102f.
degassing   107
depolymerization   226f.
deprotonation   29
DESIMS   134
desolventation
– Crown Iron "desolventizer"   216
digestion   16
diode-array detection   128, 133f., 162f., 175
dispersion energy parameter   31f.
distillation   5, 14, 16
DL-methionine   34f., 44f.
– PC-SAFT (lines)   44
DMEA   56, 58, 62f.
drug-to-extract-ration (DER)   259

e

effective diffusion coefficient   97, 99, 193, 197
EFSA, see European Food Safety Agency

ELSD, see evaporating light scattering detection
[Emim]   56, 58, 60, 62, 64f., 227
enantiomer   29f.
enzymatic hydrolysis   223, 237, 240ff.
equilibrium concentration   188f., 193, 198
equilibrium constant   29f., 50
esters   255, 257, 264
estriol   32, 34ff.
– molecular model PC-SAFT   32
estrone   34ff.
European Food Safety Agency   252, 271, 274f., 283, 286
European Medicines Agency (EMEA)   284
evaporating light scattering detection   133, 138, 162f., 166, 175
exchangers   19
– amphoteric and dipolar ion   19
– chelating ion   19
– dipolar ion   19
– selective (or specific) ion   19
extraction   5, 16
– ammonia fiber explosion (AFEX)   224
– batch   16, 195
– CO_2 recycle loop   110
– cold   5, 16
– continuous   16, 195
– conventional   88, 111f.
– counter-current solid   17
– example   191
– extraction yield   99, 102, 111
– liquid–liquid extraction   17
– microwave-assisted extraction (MAE)   186
– of organic substances   61
– oleanolic acid   80
– optimum conditions   83
– organosolv   223
– parameters   109
– pressurized fluid extraction (PFE)   186, 214
– process   149f.
– solid–liquid extraction   15
– Soxhlet   184
– steam explosion   223ff.
– system supercritical fluid extraction (SFE) 185
– technologies   64
– ultrasonic extraction (sonication)   186
– ursolic acid   80
extractor
– Basket-type   209
– BMA tower   213
– Carrousel   210
– Crown Iron Model IV   209

– De Smet LM-Extractor 211
– DSSE Screw 213
– GEA Niro Contex 212
– Lurgi sliding cell 210
– Naviglio 215
– Percolation 208
– Plug-flow 195
– Pruess Rotary 211

f
FDA 252
feed characterization 126, 129, 175
fermentation 74
fermentation process 16
flaking 100ff.
flavor 87, 89, 104
flavorings 271f.
food 269ff., 282, 285
– additive 228
– – E 406 228
– – E 407 230
– – E400 233
– – E404 233
– additives 272
– industry 273
– matrices 250, 255ff.
– supplements 270ff., 277f., 285
fucoidan 235ff.
– fucoidanase 237
– α-fucosidase 237
– structure of 237
fugacity coefficient 28, 34, 50f.
functional food 249ff., 255f., 272f., 278
– benefit areas 255
– global functional food market 273

g
gel filtration 19
gel permeation 19
Genapol 74, 78f.
General Food Law 271
GRAS 252, 257

h
health claim 272ff., 281, 286
Helmholtz energy 30ff., 49, 51ff.
hemicellulose 221ff.
[HMIM] 56, 58, 60
HTST conditioning 103, 114ff.

i
immersion 183f., 202
indicators 13
indirubin 77

international standards 281
ionic liquids 10, 223
– BMOEA bst 63
– cation and anion 57f.
– – [EMIM] 58
– DMEA oct 63
– extraction with 56
– first, second and third generation 58
– tailor-made 65
– viscosity 58f.
– water 59
irinotecan 261, 263
isoelectric point 45
isoflavone 278f., 281

k
Krafft point 68f.
Kraft pulping 224

l
labeling 270, 272ff., 278, 286
leaching 1, 15f.
lecithin 67, 71, 74
lignin 62, 221ff., 238
lignocellulose 221ff.
– biotechnological applications 224
– processing scheme for the biotransformation 222
linear solvation energy relationships LSER 13
liposome 70f.
liquid feed system 108
liquid–liquid extraction 143
– columns 154f.
– decision tree for 154
– extractors 153
– parameters 145
low density lipoprotein (LDL) 256

m
maceration 16, 183f.
market 3
– API 3f.
– nutraceuticals 3
market segments 247
mass spectrometry (MS) 133
mass transfer coefficient 97ff.
mass transfer rate 151
mass transport 99, 113
medicinal law 270f.
medicines 270
– EU legislation 270
– Traditional Herbal Medicinal Product 270

melting properties 34
membrane technology 126
mesophase 68f.
– lamellar 70
– lyotropic 68
micelle 69ff., 75
– critical micelle concentration (CMC) 68
– Formation of 67
mini-plant technology
– maceration 162
– percolation 162
– plant layout 159
– scale-down 160
– Soxhlet 161
– typical processes 164
miscibility gap 72, 74, 76
model-based experimental analysis, MEXA 199

n

natural plants production 4
natural products 6, 247
– fashionable, sales in million 249
– production chain 6
– timeline development for functional food 253
– traditional, sales in million 249
new drug applications (NDA) 259
Newtonian fluid 59
Noni-juice 277
novel food 251, 257, 269, 271f., 276f.
nutrition claim 273ff.

o

oleanoic acid 6ff.
oligosaccharide
– xylo- (XO) 225
[OMIM] 56, 58, 60
Oxycodone 264ff.
– amounts of 266
Oxycontin® 264

p

packed columns 155
paclitaxel 77, 169f., 262f.
paracetamol 40
– binary parameter 40
– PC-SAFT (lines) 41ff.
– pure-component parameters 40
partition coefficient 59, 74ff.
PC-SAFT (lines) 36ff., 41f., 47
percolation 16, 183f., 189, 202ff.
permitted daily exposure (PDE) 10

pH 29, 44f., 49
pH partitioning 60
– aniline 60
– benzoic acid 60
– thymol blue 60
pharmacopoeia 284
phase diagram 45ff., 72
– lecithin 72
– nonylphenol polyethoxylene 72
Philippines Natural Grade (PNG) 232
phyto-extraction 2
phytosterol 113ff., 256f.
piperine 191
– calibration curve 192
– extraction kinetics 193
plant ingredients
– main groups and subgroups of 127
– vs. solvent polarity 130
plant sterols 249, 256f.
plant-based extracts 143
plants 6
– from scions 6
– from sowing 6
point 48
polymorphs 33
polyphenol 113ff.
Polyphenon®E Ointment 259
polysaccharide 222ff., 226ff., 233, 235, 237f.
post-treatment 88, 107
pre-treatment 15
– of natural material 101
– pre-processing 200
– recent developments 104
– *S. lavendulifolia* 15
– solvent choice 132
press cake 102, 114ff.
press-dehydration 229f., 232
pressing 5, 15
process
– concepts 124
– design 132f.
– – cost 159
– development
– – chemical engineering approach for 128
– – general concepts for 126
– – total process development 126, 160
– – - to detailed engineering 144
– modeling 171
– – cost 175
Processed Euchema Seaweed (PES) 232

processes 106
– high pressure spray process 119
– impregnation of sprayed particles 106
– particle generation from saturated solutions (PGSS) 106
2-propanol 34, 40, 42
protonation 29
pure component 29, 50
– PC-SAFT parameters 34
pure solute 27f.
purification
– final 139, 169, 171, 174
– of latent tyrosinase 79
– pre-purification 137, 139f., 165, 171
– process scheme for Quassia 165
quality standardization 283

q
Quassia 164f.

r
racemic mandelic acid 29, 45f., 48
– PC-SAFT (lines) 47
– phase diagram 46
recovery 5, 14
– flowsheet of sporopollenin recovery 240
– of enzymes 82
– protein and myrosinase 82
– secondary metabolite 5
residue 8
– foodstuffs 9
– in artificial flavored products 9
reversed micellar (RM) extraction 75
reversed micellar solution 80ff.
rolling 100ff.
RTIL room temperature ionic liquid 57f.

s
safety assessment 283
sage 6f.
– cultivation tests 6
Salvia officinalis 6ff.
$scCO_2$ 87ff., 104ff., 111, 113, 115ff.
screw pressing 102, 116
secondary component 145, 147ff.
secondary metabolite 3, 5
secondary plant substances 278
SFE, *see* supercritical fluid extraction 89
– flow sheet 105
– of corn germ oil 116
– Patent applications 90
– plants 89
– simplified scheme 108

sieve plate 153ff., 157f.
Simulated Moving Bed 20
sitosterol 34ff.
SMB, *see* Simulated Moving Bed
solid bed 95ff., 100
solid–liquid extraction 181
– integration into flow sheet 182
– operation mode 202
solubility 1
– of selected substances 94
– point 47
– temperature dependence 38ff.
solubility data
– correlation 35, 38
– prediction 35, 42
solubility parameter 11ff.
solute 1, 11ff.
solvatochromic scales 12f.
solvent 8
– class 1 9
– class 2 10, 130
– class 3 10f., 130
– classes 33, 40, 49, 61
– co-solvent 42, 83, 94f., 107, 112
– commonly used 201
– desolventation 217
– green 60, 83
– harmless solvents 88, 95
– in food/nutraceuticals 130
– in pharmaceuticals 130
– organic 15
– other 11
– performance 147
– polarity 130f.
– recovery 62, 106, 110, 217
– selection of 14, 146, 201
– solvent power 87, 90, 94, 106
– tailor-made 83
solvent-to-solid ratio 189
Soxhlet extraction 16
special solid-state fermenters (SSF) 226
sporopollenin 227, 238ff.
– flowsheet of sporopollenin recovery 240
– structure of sporopollenin 238
spray process 118f.
standard tests 285
Sterol content 114, 116f.
Stilbenes 167
supercritical fluids
– miscibility of permanent gases in 94
– phase diagram 91
– solubility in 94
– solute concentration c 97
– tunable 93f.

supercritical fluid extraction, *see* SFE
surfactant–water-ratio 75
surfactants
– classification 66
– phosphatide 66
syneresis 229f.

t

targeted screenings 260
Taxol® 168, 179f., 262
– production flowsheet 171
Taxotere® 262
ternary 31, 47f.
terpenes 3f.
– monoterpenes 7
– triterpene 3, 7f.
thebaine 260, 264
– amounts of 266
– quantity (in tons) 265
thin layer chromography (TLC) 133f., 136
TMB, *see* true moving bed
tocopherol 116
– enrichment 116
– fractionation 117
topotecan 261, 263
trademarks 274

triterpenic acid 55, 77, 79f.
true moving bed 20
tyrosine kinase inhibitors 264

u

ursolic acid 7f.

v

valerian 104
van der Waals 30f.
Veregen® 260
Vinflunine 262

w

W0 surfactant–water-ratio 75, 80ff.
walnut 134ff.
water–acetone 42f.
water–ethanol 43f.
wellness products 250

x

xanthohumol 116ff.
xylenesulfonate 65

z

zwitterion 29
zwitterionics 65f.